AF411351

Nutrition Challenges for Middle-Aged and Older Women

Nutrition Challenges for Middle-Aged and Older Women

Editor

Masakazu Terauchi

MDPI • Basel • Beijing • Wuhan • Barcelona • Belgrade • Manchester • Tokyo • Cluj • Tianjin

Editor
Masakazu Terauchi
Department of Women's
Health, Tokyo Medical and
Dental University,
Tokyo, Japan

Editorial Office
MDPI
St. Alban-Anlage 66
4052 Basel, Switzerland

This is a reprint of articles from the Special Issue published online in the open access journal
Nutrients (ISSN 2072-6643) (available at: https://www.mdpi.com/journal/nutrients/special_issues/
Nutrition_Challenges_Middle-Aged_Older_Women).

For citation purposes, cite each article independently as indicated on the article page online and as
indicated below:

LastName, A.A.; LastName, B.B.; LastName, C.C. Article Title. *Journal Name* **Year**, *Volume Number*,
Page Range.

ISBN 978-3-0365-5477-8 (Hbk)
ISBN 978-3-0365-5478-5 (PDF)

© 2022 by the authors. Articles in this book are Open Access and distributed under the Creative
Commons Attribution (CC BY) license, which allows users to download, copy and build upon
published articles, as long as the author and publisher are properly credited, which ensures maximum
dissemination and a wider impact of our publications.
The book as a whole is distributed by MDPI under the terms and conditions of the Creative Commons
license CC BY-NC-ND.

Contents

About the Editor

Masakazu Terauchi

Masakazu Terauchi, MD, PhD, is Professor and Chair of Department of Women's Health at Tokyo Medical and Dental University. He received his medical degree in 1994 and his PhD degree in Medical Science in 2003 from Tokyo Medical and Dental University. His research interest is focused on women's health, menopause and osteoporosis. He has authored more than 300 articles in the journals Cell Metabolism, Journal of Clinical Investigation, PNAS, Menopause, Maturitas, Climacteric, etc.

He is a Certified Obstetrician-Gynecologist by the Board of Japan Society of Obstetrics and Gynecology and a North American Menopause Society Certified Menopause Practitioner.

Preface to "Nutrition Challenges for Middle-Aged and Older Women"

During menopausal transition and postmenopausal periods, women are affected by a variety of symptoms, such as hot flashes, night sweats, vaginal dryness, depression, anxiety, and insomnia. Non-specific somatic symptoms are also common, including muscle and joint pain, tiredness, and dizziness. Some of these effects (particularly vasomotor symptoms and vaginal atrophy) are closely associated with estrogen deficiency, but the exact mechanisms underlying the other symptoms are not fully understood.

Postmenopausal women are also at increased risk of cardiovascular morbidity as a net effect of central obesity, dyslipidemia, hypertension, and diabetes, as well as for osteoporosis, cognitive decline, and genitourinary syndrome of menopause.

Hormone replacement therapy (HRT) has played a central role in improving menopausal symptoms and reducing the disease risks associated with estrogen deficiency. However, due to growing concern for the side effects of HRT, especially in patients with hormone-sensitive cancer such as breast and uterus cancer, research has turned to the effects of nutraceutical approaches to these symptoms and diseases.

In this book, a variety of topics on foods, nutrients, and dietary supplements are featured, especially those that have been proved to be beneficial for the health of middle-aged and older women, including coffee, green tea, the diet of Middle-Eastern Europe, Kudzu flower–Mandarin peel, vitamin B6, chlorogenic acids, proanthocyanidins, Red Clover extracts, flavonoids, amino acids, isoflavone, etc.

I hope this book will be of help for the readers who tries to support middle-aged and older women to achieve maximum health with non-pharmacological intervention.

Masakazu Terauchi
Editor

 *nutrients*

Article

Daily Coffee and Green Tea Consumption Is Inversely Associated with Body Mass Index, Body Fat Percentage, and Cardio-Ankle Vascular Index in Middle-Aged Japanese Women: A Cross-Sectional Study

Yuka Yonekura [1], Masakazu Terauchi [2,*], Asuka Hirose [1], Tamami Odai [2], Kiyoko Kato [2] and Naoyuki Miyasaka [1]

[1] Department of Obstetrics and Gynecology, Tokyo Medical and Dental University, Yushima 1-5-45, Bunkyo, Tokyo 113-8510, Japan; 160992ms@tmd.ac.jp (Y.Y.); a-kacrm@tmd.ac.jp (A.H.); n.miyasaka.gyne@tmd.ac.jp (N.M.)

[2] Department of Women's Health, Tokyo Medical and Dental University, Yushima 1-5-45, Bunkyo, Tokyo 113-8510, Japan; odycrm@tmd.ac.jp (T.O.); kiyo.crm@tmd.ac.jp (K.K.)

* Correspondence: teragyne@tmd.ac.jp; Tel./Fax: +81-3-5803-4605

Received: 26 March 2020; Accepted: 7 May 2020; Published: 11 May 2020

Abstract: This study aimed to investigate the links between coffee (CF)/green tea (GT) consumption and body composition/cardiovascular parameters in middle-aged Japanese women. We conducted a cross-sectional study of 232 Japanese women aged 40–65 years who had been referred to the menopause clinic of Tokyo Medical and Dental University Hospital between November 2007 and August 2017. Body composition, cardiovascular parameters, and CF/GT consumption frequency were evaluated on their initial visits, using a body composition analyzer, vascular screening system, and brief-type self-administered diet history questionnaire, respectively. We investigated the associations between variables using multivariate logistic regression. After adjustment for age, menopausal status, and other factors, daily CF consumption was inversely associated with high body mass index (BMI) (adjusted odds ratio, 0.14; 95% confidence interval, 0.14–0.96) and body fat percentage (BF%) (0.33; 0.14–0.82), and daily GT consumption with high BF% (0.36; 0.14–0.96). Daily CF + GT consumption was also inversely associated with high BMI (0.15; 0.05–0.50) and BF% (0.30; 0.12–0.74). In pre- and perimenopausal women, daily CF + GT consumption was inversely associated with high cardio-ankle vascular index (CAVI) (0.05; 0.003–0.743). In conclusion, daily CF/GT consumption was inversely associated with high BMI, BF%, and CAVI in middle-aged Japanese women.

Keywords: menopause; atheromatous arteriosclerosis; metabolic syndrome; obesity; caffeine; polyphenol

---

1. Introduction

In the menopausal transition (perimenopausal) and postmenopausal periods, a reduction in estrogen levels not only causes various symptoms such as hot flashes, night sweats, vaginal dryness, depression, anxiety, and insomnia, but also changes fat distribution from subcutaneous to visceral adiposity, which is one of the major risk factors of atheromatous arteriosclerosis. Atherosclerosis is a chronic inflammatory disease [1]. It is the main pathophysiological cause of cardiovascular diseases (CVDs), which are the biggest causes of death worldwide. According to the World Health Organization (WHO), 17.9 million people died of CVDs in 2016 [2]. Considering that the prevalence of CVDs steeply increases after menopause [3–6], it is important to lower the risk of atheromatous arteriosclerosis in middle-aged women.

Healthy lifestyles, including a well-balanced diet, physical activity, quitting smoking, and alcohol moderation, are well-known preventive measures of atherosclerosis. It has also been reported that beverages such as coffee (CF) and green tea (GT) are effective in lowering cardiovascular risks. A couple of human trials have revealed the anti-obesity effects of CF and GT extracts [7,8]. CF and GT consumption is also associated with a lower risk of stroke [9]. Furthermore, a pilot study reported that CF extract improves arterial stiffness in healthy Japanese men [10], and a meta-analysis showed that GT intake improves cardiovascular risk markers, including systolic blood pressure [11]. However, the effects of CF and GT intakes in middle-aged women, who are threatened by increasing cardiovascular risks, have yet to be elucidated. Cardio-ankle vascular index (CAVI) is one of the noninvasive measures used as an index of atherosclerosis. CAVI is superior to other measures in that we could also measure other cardiovascular parameters at the same time, and it could well represent the arterial stiffness from the origin of aorta to ankle [12–14]. The present study aimed to investigate the associations between CF/GT consumption and body composition/cardiovascular parameters, including CAVI, in middle-aged Japanese women.

2. Materials and Methods

2.1. Study Population

In this cross-sectional study, we analyzed the first-visit records of 232 Japanese women who were enrolled in the Systematic Health and Nutrition Education Program at the menopause clinic of Tokyo Medical and Dental University Hospital, Japan, from November 2007 to August 2017. All women who enrolled in this program had attended our clinic to treat menopausal symptoms. The collected data included age, menopausal status, body composition parameters, cardiovascular parameters, lifestyle factors, and detailed dietary habits. These data were collected by physicians and nutritionists on their initial visits. The inclusion criteria were those aged between 40 and 65 years, reported their menopausal status, and completed the brief-type self-administered diet history questionnaire (BDHQ). The exclusion criteria were those who had been treated with menopausal hormone therapy, antihyperlipidemic drugs, and antidiabetic agents, whose body mass index (BMI) was <16 kg/m^2 or >35 kg/m^2, and who answered in the BDHQ that they drank black tea ≥ 1 cup/day.

The research protocol was reviewed and approved by the Tokyo Medical and Dental University Review Board (approval number: 774), and written informed consent was obtained from all participants. The study was conducted in accordance with the Declaration of Helsinki.

2.2. Measurements

2.2.1. Menopausal Status

Participants were classified as premenopausal if they had regular menstrual cycles; perimenopausal if they had a menstrual period within the past 12 months but had missed periods or had an irregular cycle in the past 3 months; and postmenopausal if they had no menstruation in the past 12 months.

2.2.2. Physical Assessments and Cardiovascular Parameters

Body composition parameters, including height, weight, BMI, fat mass, muscle mass, and lean body mass, were measured using a body composition analyzer (MC190-EM; Tanita, Tokyo, Japan). Resting energy expenditure was calculated based on respiratory volume using a portable indirect calorimeter (Metavine-N VMB-005 N; Vine, Tokyo, Japan). Additionally, cardiovascular parameters, including systolic and diastolic blood pressure, heart rate, cardio-ankle vascular index (CAVI), and ankle-brachial pressure index (ABI), were assessed using a vascular screening system (VS-1000; Fukuda Denshi, Tokyo, Japan). Plasma blood sugar level was also measured. Blood examination was conducted in accordance with the guidelines on internal and external quality control defined by the Japanese Ministry of Health, Labor, and Welfare.

2.2.3. Lifestyle Characteristics

Participants underwent a medical interview for lifestyle factors, which included the frequency of alcohol consumption (never, sometimes, every day), smoking (no, yes), and regular exercise habits (none, <3 times/week, ≥3 times/week).

2.2.4. Dietary Habits

Dietary habits were assessed using the BDHQ. The BDHQ asked about the consumption frequencies of selected food and beverage items commonly consumed in Japan. Based on the participants' responses to the questionnaire, an ad-hoc computer algorithm estimated the amounts of 96 nutrients consumed during the previous month. The choices for the frequencies of CF and GT consumption were as follows: 0, <1, 1, 2–3, or 4–6 times/week, or 1, 2–3, or ≥4 cups/day. Energy intake was calculated based on the responses to the BDHQ.

2.3. Factors Associated with Body Composition/Cardiovascular Parameters

We dichotomized CF and GT consumption as low (<1 cup/day) and high (≥1 cup/day), and the women were divided into four groups according to their consumption frequency; (i) control group, who consumed < 1 cup/day of both CF and GT; (ii) CF group, who consumed ≥ 1 cup/day of CF alone; (iii) GT group, who consumed ≥ 1 cup/day of GT alone; and (iv) CF + GT group, who consumed ≥ 1 cup/day of both CF and GT. To determine the background factors which could be confounding, women in these four groups were compared for background factors, including age, menopausal status, height, basal metabolism, energy intake, and lifestyle characteristics. The factors that significantly differed ($p < 0.05$) among the groups were selected as explanatory variables for a multivariate logistic regression analysis.

2.4. Statistical Analysis

The required total sample size was estimated at 233, as calculated from the number of predictive variables, events per variable, and event incidence rate of 7, 10, and 0.30, respectively. First, to determine which parameters were associated with CF and GT consumption, we compared body composition and cardiovascular parameters among four groups. Next, we conducted a multivariate logistic regression analysis to find out whether CF/GT consumption was independently associated with the selected body composition/cardiovascular parameters. We examined the association, adjusting for the extracted background factors. When CF/GT consumption frequency retained a significant association ($p < 0.05$) with the selected body composition/cardiovascular parameters in the final multivariate model, we considered that CF/GT consumption was associated with the selected body composition/cardiovascular parameters in Japanese middle-aged women. Statistical analysis, including a one-way analysis of variance (ANOVA) with Tukey's multiple comparison test and a chi-squared test, was performed with GraphPad Prism version 5.02 (GraphPad Software, San Diego, CA, USA). The multivariate logistic regression analysis was performed with JMP version 14 (SAS Institute Inc., Cary, NC, USA). A *p*-value of <0.05 was considered statistically significant.

3. Results

The average age of the participants was 51.6 ± 5.0 years (mean ± standard deviation). The number of women who were classified into the control, CF, GT, and CF + GT groups was 39 (16.8%), 47 (20.3%), 76 (32.8%), and 70 (30.2%), respectively.

We first compared body composition, cardiovascular parameters, and background characteristics among the four groups using a one-way ANOVA with Tukey's multiple comparison test and a chi-squared test. The body composition and cardiovascular parameters that statistically differed ($p < 0.05$) at the univariate level among the four groups were height (cm), body weight (kg), BMI (kg/m^2), body fat (%), and fat mass (kg) (Table 1).

Table 1. Comparison of background characteristics and daily consumption of coffee (CF) and green tea (GT).

	Control (n = 39)	CF (n = 76)	GT (n = 47)	CF + GT (n = 70)	*p*-Value
Age and menopausal status					
Age (years), mean (SD)	51.2 (6.1)	50.4 (3.8)	52.0 (5.2)	52.8 (5.3)	0.027 [a]
Menopausal status					
Premenopause	11 (28.2)	31 (40.8)	7 (14.9)	17 (24.3)	
Perimenopause	8 (20.5)	20 (26.3)	6 (12.8)	11 (15.7)	
Postmenopause	20 (51.3)	25 (32.9)	34 (72.3)	42 (60.0)	0.002 [b]
Body composition, mean (SD)					
Height (cm)	157.8 (6.9)	158.7 (5.1)	155.5 (6.4)	157.6 (4.9)	0.028 [a]
Body weight (kg)	57.4 (12.7)	53.9 (8.1)	52.4 (9.6)	52.1 (7.1)	0.022 [a]
Body mass index (kg/m^2)	23.0 (4.4)	21.4 (3.2)	21.6 (3.5)	21.0 (2.7)	0.027 [a]
Body fat (%)	30.5 (8.5)	26.8 (7.0)	27.2 (7.5)	25.8 (6.6)	0.014 [a]
Fat mass (kg)	18.5 (8.9)	14.9 (6.2)	14.8 (6.7)	13.8 (5.1)	0.005 [a]
Muscle mass (kg)	36.7 (4.2)	36.7 (2.8)	35.4 (3.7)	36.1 (2.9)	0.140 [a]
Lean body mass (kg)	38.9 (4.5)	39.0 (3.1)	37.6 (4.0)	38.3 (3.2)	0.140 [a]
Cardiovascular parameters, mean (SD)					
Systolic blood pressure (mmHg)	131 (21)	126 (18)	126 (20)	125 (16)	0.371 [a]
Diastolic blood pressure (mmHg)	83 (14)	80 (11)	81 (13)	80 (11)	0.575 [a]
Pulse rate (beats/minute)	66 (16)	64 (11)	64 (11)	64 (11)	0.822 [a]
Blood sugar level (mg/dL)	98 (23)	96 (8)	100 (14)	95 (8.4)	0.319 [a]
Cardio-ankle vascular index	7.78 (0.84)	7.45 (0.70)	7.54 (0.65)	7.50 (0.71)	0.178 [a]
Ankle-brachial pressure index	1.11 (0.06)	1.11 (0.06)	1.12 (0.07)	1.10 (0.05)	0.475 [a]
Basal metabolism, mean (SD)					
Resting energy expenditure (kcal/day)	1669 (527)	1628 (385)	1618 (419)	1512 (393)	0.220 [a]
Dietary intake, mean (SD)					
Energy intake (kcal/day)	1490 (532)	1638 (394)	1735 (544)	1786 (468)	0.013 [a]
Lifestyle characteristics, number (%)					
Smoking habit					
Yes	2 (5.1)	10 (13.2)	0 (0)	4 (5.7)	
No	37 (94.9)	66 (86.8)	47 (100)	66 (94.3)	0.037 [b]
Frequency of alcohol consumption					
Every day	3 (7.7)	13 (17.1)	7 (14.9)	2 (2.9)	
Sometimes	14 (35.9)	28 (36.8)	8 (17.0)	21 (30.0)	
Nondrinker	22 (56.4)	35 (46.1)	32 (68.1)	47 (67.1)	0.015 [b]
Regular exercise habit					
≥3 times/week	1 (2.6)	8 (10.5)	6 (12.8)	5 (7.1)	
1–2 times/week	18 (46.2)	13 (17.1)	12 (25.5)	18 (25.7)	
None	20 (51.3)	55 (72.4)	29 (61.7)	47 (67.1)	0.039 [b]

[a] One-way analysis of variance. [b] Chi-squared test. CF, coffee; GT, green tea; SD, standard deviation.

Of these parameters, body weight, BMI, body fat percentage (BF%), and fat mass showed a similar decreasing tendency in the order of control > CF, GT > CF + GT. Considering that body weight and fat mass were dependent on height, we decided to limit further analysis of the associations between BMI/BF% and daily CF/GT consumption. Regarding background characteristics related to BMI and BF%, we selected age, menopausal status, energy intake, smoking habit, frequency of alcohol consumption, and regular exercise habit, which statistically differed among the four groups ($p < 0.05$, Table 1). We performed multivariate logistic regression analysis to determine the independent relationships between daily CF/GT consumption and high BMI (≥25 kg/m^2)/high BF% (≥30%).

After adjustment for age and menopausal status (Model 2), and also for selected background characteristics (Model 3), both CF consumption and CF + GT consumption of > 1 cup/day exhibited a significant inverse relationship with high BMI (CF: adjusted odds ratio (OR], 0.14; 95% confidence interval (CI], 0.05–0.46; $p < 0.01$; and CF + GT: OR, 0.15; 95% CI, 0.05–0.50; $p < 0.01$), while the consumption of all three, CF, GT, and CF + GT, was inversely associated with high BF% (CF: OR, 0.33; 95% CI, 0.14–0.82; $p < 0.05$; GT: OR, 0.36; 95% CI, 0.14–0.96; $p < 0.05$; and CF + GT: OR, 0.30; 95% CI, 0.12–0.74; $p < 0.01$) (Table 2).

Table 2. Multivariate analysis of the associations of daily coffee (CF) and green tea (GT) consumption with high body mass index (BMI) and body fat percentage (BF%).

Model	Group	BMI ≥ 25			BF% ≥ 30		
		OR	95% CI	*p*-Value	OR	95% CI	*p*-Value
Model 1	CF	0.26	0.10–0.72	<0.01	0.42	0.19–0.93	<0.05
	GT	0.53	0.20–1.44	0.2	0.49	0.20–1.15	0.1
	CF + GT	0.21	0.07–0.62	<0.01	0.34	0.15–0.78	<0.05
Model 2	CF	0.25	0.09–0.69	<0.01	0.43	0.19–0.95	<0.05
	GT	0.57	0.21–1.56	0.3	0.46	0.19–1.12	0.09
	CF + GT	0.21	0.07–0.64	<0.01	0.35	0.15–0.80	<0.05
Model 3	CF	0.14	0.05–0.46	<0.01	0.33	0.14–0.82	<0.05
	GT	0.38	0.12–1.18	0.09	0.36	0.14–0.96	<0.05
	CF + GT	0.15	0.05–0.50	<0.01	0.30	0.12–0.74	<0.01

BMI, body mass index; BF%, body fat percentage; CI, confidence interval; OR, odds ratio CF, coffee; GT, green tea.

We also conducted subgroup analyses of the associations between CF/GT consumption and body composition/cardiovascular parameters in pre-/perimenopausal and postmenopausal women in the same manner. At the univariate level, a significant difference was not observed with regard to body composition. As for cardiovascular parameters, CAVI significantly differed among the four groups in pre-/perimenopausal women and exhibited a decreasing trend. Using CAVI = 8.0 as the cutoff value, we performed multivariate logistic regression analysis to investigate the relationship between daily CF/GT consumption and high CAVI (≥8.0). After adjustment for age (Model 2), and also for selected background characteristics (Model 3), daily CF + GT consumption was inversely associated with high CAVI (OR, 0.05; 95% CI, 0.003–0.743; $p < 0.05$) (Table 3).

Table 3. Multivariate analysis of the associations of daily coffee (CF) and green tea (GT) consumption with high cardio-ankle vascular index (CAVI) in pre-/perimenopausal women.

Model	Group	CAVI ≥ 8.0		
		OR	95% CI	*p*-Value
Model 1	CF	0.47	0.12–1.94	0.3
	GT	0.23	0.02–2.39	0.2
	CF + GT	0.21	0.03–1.33	0.1
Model 2	CF	0.38	0.08–1.91	0.2
	GT	0.13	0.01–1.66	0.1
	CF + GT	0.12	0.02–1.00	<0.05
Model 3	CF	0.37	0.05–2.53	0.3
	GT	0.11	0.01–2.27	0.2
	CF + GT	0.05	0.003–0.743	<0.05

CAVI, cardio-ankle vascular index; CI, confidence interval; OR, odds ratio; CF, coffee; GT, green tea.

4. Discussion

In this cross-sectional analysis of 232 middle-aged Japanese women who attended our menopause clinic, the daily intake of CF/GT was shown to be inversely associated with high BMI and BF%. Moreover, in pre- and perimenopausal women, the daily intake of both CF and GT was inversely associated with high CAVI.

CF comprises many components with pharmacologic effects, such as caffeine and chlorogenic acid (CGA). CGA is one of the polyphenols that are abundant in green CF beans and is considered to be the most active compound [15–17]. Some studies showed that CGA has the effect of suppressing the accumulation of body fat. A randomized, double-blind trial showed that the daily consumption of CF rich in CGA for 12 weeks by overweight Japanese adults lowered the visceral fat area, total abdominal fat area, BMI, and waist circumference [7]. It was also reported that coffee reduced the accumulation of lipids during adipocytic differentiation of 3T3-L1 preadipocytes and inhibited the expression of peroxisome proliferator-activated receptor γ, which controls the differentiation of adipocytes [18]. Another underlying mechanism was suggested in human trials. These trials reported that CGA consumption increased postprandial energy expenditure and fat utilization in healthy humans [19,20]. It is plausible that CF ingredients may promote fat oxidation and suppress fat differentiation in humans.

In addition, CGA also has anti-atherosclerotic properties [21]. A single-blind, randomized, placebo-controlled, crossover trial showed that a single intake of CF with a high content of CGA and a low content of hydroxyhydroquinone, which oxidizes CGA and reduces its function, improved postprandial endothelial dysfunction by decreasing oxidative stress [22]. A recent meta-analysis of randomized clinical trials reported the antihypertensive effect of CGA [23]. Moreover, a randomized controlled trial showed that coffee rich in caffeoylquinic acid, one of the CGAs, improved the levels of plasma total cholesterol, triglycerides, and low-density lipoprotein cholesterol [24]. These results indicate that CGA could be effective in preventing atherosclerosis.

GT also contains several green tea catechins (GTCs), such as epicatechin, epicatechin gallate, epigallocatechin, and epigallocatechin gallate (EGCG), as well as their thermal isomers, such as catechin, catechin gallate, gallocatechin, and gallocatechin gallate [25]. Of these catechins, EGCG is the most abundant in GT infusions and is considered to be the most active compound [25,26]. A pooled analysis of six human trials demonstrated that the consumption of beverages containing GTCs, mainly EGCG, for 12 weeks significantly reduced the total fat area, visceral fat area, BMI, and waist circumference [27]. A cell-based experiment partly revealed the mechanism with which EGCG inhibits 3T3-L1 preadipocyte differentiation, activates AMPK, and decreases fat accumulation [28]. GTCs could also have protective effects on endothelial cells as other antioxidants do, by inhibiting the adhesion of monocytes [29,30]. In another cell-based study, EGCG was shown to suppress the mRNA expression of monocyte chemotactic protein-1, which accelerates the progress of atherosclerosis by promoting the adhesion of monocytes [31]. These findings indicate that GT may be a potential therapeutic agent for the prevention of atherosclerosis.

In this study, the combined intake of CF and GT was inversely associated with high CAVI in pre- and perimenopausal women; however, the intake of CF or GT alone did not show any inverse relationship. These results suggest the additive effect of CF and GT in lowering CAVI. A large cohort study in Japan showed that the consumption of ≥4 cups/day of GT and a combined intake of ≥1 cup/day of CF and ≥2 cups/day of GT contributed to a risk reduction in CVD and stroke [11]. Considering that a cup of GT contains about 112 mg EGCG and a cup of CF contains about 160 mg CGA [32], it is possible that a daily intake of 400 mg of polyphenols is effective in preventing CVD. Although the mechanism underlying the combined effect remains unclear, the different antioxidants in CF and GT may additively strengthen the beneficial effects on body composition and cardiovascular parameters. In another population-based prospective study in Japan, the researchers found that dietary polyphenol intake is inversely associated with CVD [33]. To corroborate the association between polyphenols and CVD risk, more comprehensive studies are warranted.

In this study, the daily intake of CF/GT was not inversely associated with high CAVI in postmenopausal women. After menopause, a reduction in estrogen levels causes rapid endothelial dysfunction, which could not be overcome by the moderate effects of CF and GT on vascular endothelial function.

This study has several limitations. First, the study population was relatively small and consisted only of middle-aged Japanese women who attended our menopause clinic. Therefore, generalizing our results to a wider population is difficult. Second, we did not consider polyphenols from foods other than CF and GT in this study. Finally, the cross-sectional design of this study prevented the determination of a causal relationship between the daily intake of CF/GT and lower BMI, BF%, and CAVI.

Nevertheless, this study has several strengths. Background factors associated with body composition and cardiovascular parameters were well investigated, including energy intake and expenditure, and lifestyles. To corroborate the observed association between the daily intake of CF/GT and body composition, we recommend prospective studies that enroll a higher number of obese and hypertensive women and evaluate both daily intake and serum levels of the polyphenols.

5. Conclusions

This study revealed that daily CF and GT consumption was inversely associated with high BMI and BF% in middle-aged Japanese women. In pre- and perimenopausal women, the daily consumption of both CF and GT was inversely associated with high CAVI. These results suggest that the consumption of CF and GT could help women to keep fit and prevent atherosclerosis.

Author Contributions: Y.Y., M.T., and T.O.; were responsible for project development, data collection, and data analysis. K.K. was responsible for data collection. A.H. and N.M. were responsible for project development and supervision. All authors contributed to reviewing and editing the manuscript. All authors have read and agreed to the published version of the manuscript.

Funding: This research was funded by the Honjo International Scholarship Foundation (X2977).

Conflicts of Interest: The authors declare no conflict of interest. The funders had no role in the design of the study; in the collection, analyses, or interpretation of data; in the writing of the manuscript; or in the decision to publish the results.

References

1. Ross, R. Atherosclerosis—An inflammatory disease. *N. Engl. J. Med.* **1999**, *340*, 115–126. [CrossRef] [PubMed]
2. World Health Organization. Cardiovascular Diseases (CVDs). Available online: https://www.who.int/news-room/fact-sheets/detail/cardiovascular-diseases-(cvds) (accessed on 18 October 2019).
3. Carr, M.C. The emergence of the metabolic syndrome with menopause. *J. Clin. Endocrinol. Metab.* **2003**, *88*, 2404–2411. [CrossRef] [PubMed]
4. Rosano, G.M.; Vitale, C.; Tulli, A. Managing cardiovascular risk in menopausal women. *Climacteric* **2006**, *9*, 19–27. [CrossRef] [PubMed]
5. Lizcano, F.; Guzmán, G. Estrogen deficiency and the origin of obesity during menopause. *BioMed Res. Int.* **2014**, *2014*, 757461. [CrossRef]
6. Collins, P.; Rosano, G.; Casey, C.; Daly, C.; Gambacciani, M.; Hadji, P.; Kaaja, R.; Mikkola, T.; Palacios, S.; Preston, R.; et al. Management of cardiovascular risk in the peri-menopausal woman: A consensus statement of European cardiologists and gynaecologists. *Eur. Heart J.* **2007**, *28*, 2028–2040. [CrossRef]
7. Watanabe, T.; Kobayashi, S.; Yamaguchi, T.; Hibi, M.; Fukuhara, I.; Osaki, N. Coffee abundant in chlorogenic acids reduces abdominal fat in overweight adults: A randomized, double-blind, controlled trial. *Nutrients* **2019**, *11*, 1617. [CrossRef]
8. Nagao, T.; Hase, T.; Tokimitsu, I. A green tea extract high in catechins reduces body fat and cardiovascular risks in humans. *Obesity* **2007**, *15*, 1473–1483. [CrossRef]
9. Kokubo, Y.; Iso, H.; Saito, I.; Yamagishi, K.; Yatsuya, H.; Ishihara, J.; Inoue, M.; Tsugane, S. The impact of green tea and coffee consumption on the reduced risk of stroke incidence in Japanese population: The Japan public health center-based study cohort. *Stroke* **2013**, *44*, 1369–1374. [CrossRef]
10. Suzuki, A.; Nomura, T.; Jokura, H.; Kitamura, N.; Saiki, A.; Fujii, A. Chlorogenic acid-enriched green coffee bean extract affects arterial stiffness assessed by the cardio-ankle vascular index in healthy men: A pilot study. *Int. J. Food Sci. Nutr.* **2019**, *70*, 901–908. [CrossRef]
11. Khalesi, S.; Sun, J.; Buys, N.; Jamshidi, A.; Nikbakht-Nasrabadi, E.; Khosravi-Boroujeni, H. Green tea catechins and blood pressure: A systematic review and meta-analysis of randomised controlled trials. *Eur. J. Nutr.* **2014**, *53*, 1299–1311. [CrossRef]
12. Parsanathan, R.; Jain, S.K. Novel Invasive and Noninvasive Cardiac-Specific Biomarkers in Obesity and Cardiovascular Diseases. *Metab. Syndr. Relat. Disord.* **2020**, *18*, 10–30. [CrossRef] [PubMed]
13. Saiki, A.; Sato, Y.; Watanabe, Y.; Imamura, H.; Yamaguchi, T.; Ban, N.; Kawana, H.; Nagumo, A.; Nagayama, D.; Ohira, M.; et al. The Role of a Novel Arterial Stiffness Parameter, Cardio-Ankle Vascular Index (CAVI), as a Surrogate Marker for Cardiovascular Diseases. *J. Atheroscler. Thromb.* **2016**, *23*, 155–168. [CrossRef] [PubMed]

14. Shirai, K.; Saiki, A.; Nagayama, D.; Tatsuno, I.; Shimizu, K.; Takahashi, M. The Role of Monitoring Arterial Stiffness with Cardio-Ankle Vascular Index in the Control of Lifestyle-Related Diseases. *Pulse* **2015**, *3*, 118–133. [CrossRef] [PubMed]

15. Gonzalez de Mejia, E.; Ramirez-Mares, M.V. Impact of caffeine and coffee on our health. *Trends Endocrinol. Metab.* **2014**, *25*, 489–492. [CrossRef]

16. Martini, D.; Del Bo, C.; Tassotti, M.; Riso, P.; Del Rio, D.; Brighenti, F.; Porrini, M. Coffee consumption and oxidative stress: A review of human intervention studies. *Molecules* **2016**, *21*, 979. [CrossRef]

17. Fukushima, Y.; Ohie, T.; Yonekawa, Y.; Yonemoto, K.; Aizawa, H.; Mori, Y.; Watanabe, M.; Takeuchi, M.; Hasegawa, M.; Taguchi, C.; et al. Coffee and green tea as a large source of antioxidant polyphenols in the Japanese population. *J. Agric. Food Chem.* **2009**, *57*, 1253–1259. [CrossRef]

18. Aoyagi, R.; Funakoshi-Tago, M.; Fujiwara, Y.; Tamura, H. Coffee inhibits adipocyte differentiation via inactivation of PPARγ. *Biol. Pharm. Bull.* **2014**, *37*, 1820–1825. [CrossRef]

19. Soga, S.; Ota, N.; Shimotoyodome, A. Stimulation of postprandial fat utilization in healthy humans by daily consumption of chlorogenic acids. *Biosci. Biotechnol. Biochem.* **2014**, *77*, 1633–1636. [CrossRef]

20. Ota, N.; Soga, S.; Murase, T.; Shimotoyodome, A.; Hase, T. Consumption of coffee polyphenols increases fat utilization in humans. *J. Health Sci.* **2010**, *56*, 745–751. [CrossRef]

21. Yun, N.; Kang, J.W.; Lee, S.M. Protective effects of chlorogenic acid against ischemia/reperfusion injury in rat liver: Molecular evidence of its antioxidant and anti-inflammatory properties. *J. Nutr. Biochem.* **2012**, *23*, 1249–1255. [CrossRef]

22. Kajikawa, M.; Maruhashi, T.; Hidaka, T.; Nakano, Y.; Kurisu, S.; Matsumoto, T.; Iwamoto, Y.; Kishimoto, S.; Matsui, S.; Aibara, Y.; et al. Coffee with a high content of chlorogenic acids and low content of hydroxyhydroquinone improves postprandial endothelial dysfunction in patients with borderline and stage 1 hypertension. *Eur. J. Nutr.* **2019**, *58*, 989–996. [CrossRef]

23. Onakpoya, I.J.; Spencer, E.A.; Thompson, M.J.; Heneghan, C.J. The effect of chlorogenic acid on blood pressure: A systematic review and meta-analysis of randomized clinical trials. *J. Hum. Hypertens.* **2015**, *29*, 77–81. [CrossRef] [PubMed]

24. Martínez-López, S.; Sarriá, B.; Mateos, R.; Bravo-Clemente, L. Moderate consumption of a soluble green/roasted coffee rich in caffeoylquinic acids reduces cardiovascular risk markers: Results from a randomized, cross-over, controlled trial in healthy and hypercholesterolemic subjects. *Eur. J. Nutr.* **2019**, *58*, 865–878. [CrossRef] [PubMed]

25. Del Rio, D.; Stewart, A.J.; Mullen, W.; Burns, J.; Lean, M.E.; Brighenti, F.; Crozier, A. HPLC-MSn analysis of phenolic compounds and purine alkaloids in green and black tea. *J. Agric. Food Chem.* **2004**, *52*, 2807–2815. [CrossRef] [PubMed]

26. Higdon, J.V.; Frei, B. Tea catechins and polyphenols: Health effects, metabolism, and antioxidant functions. *Crit. Rev. Food Sci. Nutr.* **2003**, *43*, 89–143. [CrossRef] [PubMed]

27. Hibi, M.; Takase, H.; Iwasaki, M.; Osaki, N.; Katsuragi, Y. Efficacy of tea catechin-rich beverages to reduce abdominal adiposity and metabolic syndrome risks in obese and overweight subjects: A pooled analysis of 6 human trials. *Nutr. Res.* **2018**, *55*, 1–10. [CrossRef]

28. Moon, H.S.; Chung, C.S.; Lee, H.G.; Kim, T.G.; Choi, Y.J.; Cho, C.S. Inhibitory effect of (-)-epigallocatechin-3-gallate on lipid accumulation of 3T3-L1 cells. *Obesity* **2007**, *15*, 2571–2582. [CrossRef]

29. Parsanathan, R.; Jain, S.K. L-Cysteine in vitro can restore cellular glutathione and inhibits the expression of cell adhesion molecules in G6PD-deficient monocytes. *Amino Acids* **2018**, *50*, 909–921. [CrossRef]

30. Parsanathan, R.; Jain, S.K. Glucose-6-phosphate dehydrogenase deficiency increases cell adhesion molecules and activates human monocyte-endothelial cell adhesion: Protective role of l-cysteine. *Arch. Biochem. Biophys.* **2019**, *663*, 11–21. [CrossRef]

31. Li, Y.F.; Wang, H.; Fan, Y.; Shi, H.J.; Wang, Q.M.; Chen, B.R.; Khurwolah, M.R.; Long, Q.Q.; Wang, S.B.; Wang, Z.M.; et al. Epigallocatechin-3-Gallate Inhibits Matrix Metalloproteinase-9 and Monocyte Chemotactic Protein-1 Expression Through the 67-κDa Laminin Receptor and the TLR4/MAPK/NF-κB Signalling Pathway in Lipopolysaccharide-Induced Macrophages. *Cell. Physiol. Biochem.* **2017**, *43*, 926–936. [CrossRef]

32. Williamson, G.; Dionisi, F.; Renouf, M. Flavanols from green tea and phenolic acids from coffee: Critical quantitative evaluation of the pharmacokinetic data in humans after consumption of single doses of beverages. *Mol. Nutr. Food Res.* **2011**, *55*, 864–873. [CrossRef] [PubMed]

33. Taguchi, C.; Kishimoto, Y.; Fukushima, Y.; Kondo, K.; Yamakawa, M.; Wada, K.; Nagata, C. Dietary intake of total polyphenols and the risk of all-cause and specific-cause mortality in Japanese adults: The Takayama study. *Eur. J. Nutr.* **2020**, *59*, 1263–1271. [CrossRef] [PubMed]

© 2020 by the authors. Licensee MDPI, Basel, Switzerland. This article is an open access article distributed under the terms and conditions of the Creative Commons Attribution (CC BY) license (http://creativecommons.org/licenses/by/4.0/).

Article

The Change in the Content of Nutrients in Diets Eliminating Products of Animal Origin in Comparison to a Regular Diet from the Area of Middle-Eastern Europe

Kamila Kowalska [1], Jacek Brodowski [2], Kamila Pokorska-Niewiada [3] and Małgorzata Szczuko [1,*

[1] Department of Human Nutrition and Metabolomics, Pomeranian Medical University in Szczecin, Broniewskiego 24, 71-460 Szczecin, Poland; mila1994kow@gmail.com

[2] Primary Care Department, Pomeranian Medical University in Szczecin, Żołnierska 48, 71-210 Szczecin, Poland; jacek.brodowski@pum.edu.pl

[3] Department of Toxicology, Dairy Technology and Food Storage, West Pomeranian University of Technology in Szczecin, Papieża Pawła VI 3, 71-459 Szczecin, Poland; kamila.pokorska@zut.edu.pl

* Correspondence: malgorzata.szczuko@pum.edu.pl; Tel.: +48-91-441-4810; Fax: +48-91-441-4807

Received: 27 August 2020; Accepted: 25 September 2020; Published: 29 September 2020

Abstract: Introduction: The diet of Poles became similar to the western style of nutrition. It is rich in saturated fats, it contains significant quantities of salt, and has very low fruit and vegetable content. On the other hand, introducing an incorrectly planned diet that eliminates animal products may be associated with the risk of deficiencies of certain vitamins and minerals. Taking into account the regular diet of Poles, a properly balanced vegetarian menu may be a better and safer choice for the proper functioning of the organism. Aim: The analysis of the content of individual types of vegetarian diets and a comparison with the menus of the regular diet of the Polish population. Materials and methods: 70 menus were subjected to a quantitative analysis, 10 menus for each 7 type of diet eliminating products of animal origin and regular diets without elimination. The caloricity of the designed diets was ±2000 kcal. The quantitative evaluation of the menus was performed using the Dieta 6d dietary program. Statistical significance was established at $p \leq 0.05$. Results: It was observed that the regular diet of Poles (RD) featured the highest content of total fats, as well as saturated acids and cholesterol. The VEGAN diet was characterized by the lowest total protein content and the lack of wholesome protein and cholesterol. RD was characterized by the lowest average content of dietary fiber. The highest content of saccharose was observed in RD. Sodium content in RD significantly exceeded the recommended daily norm. RD featured insufficient content of the following minerals and vitamins: potassium, calcium, magnesium, iodine, Vitamin E, Vitamin C, folates, and Vitamin D. The norm for calcium has not been fulfilled also in milk-free and vegan diets. All of the analyzed diets lacked proper amounts of iodine and Vitamin D. The highest content of polyunsaturated fatty acids was observed in the VEGAN diet. The periodic elimination of meat and fatty dairy products should be included in the treatment of the metabolic syndrome, hypertensions, hyperlipidemia, obesity, and type 2 diabetes. Conclusions: The regular diet of Poles turned out to be more dangerous for health in terms of deficiencies than properly balanced diets eliminating products of animal origin.

Keywords: vegetarian diets; vegetarianism; vitamins; minerals; nutritional habits of Poles

---

1. Introduction

Humans are anatomically and physiologically fit to consume products both of plant and animal origin [1]. However, the current trends associated with mass production often result in people resigning from the consumption of meat and other products of animal origin. Other reasons for the elimination of such products from the diet include religious beliefs, views relating to health and ecology, as well as the economic situation [2]. Every type of diet that eliminates animal products should fulfill the need of an individual for nutrients, and in the case of their deficiencies, it is important to include supplementation. Properly balanced diets eliminating products of animal origin are not a threat to health. However, during the period of growth, pregnancy, or lactation, and in sportsmen, significant nutritional restrictions in the form of veganism may entail the risk of numerous nutritional deficiencies and illnesses (having in mind incorrectly planned diets). Despite the growing interest of people in vegetarian diets, the average consumption of meat in developed countries is still at a level that is too high, especially among men [3]. The current information on public health recommend the consumption of up to three portions of red meat (350g–500g per week), whereas processed meat products should be avoided or limited as much as possible [4]. Women more often become vegetarians than men [3]. Lower body mass index (BMI), cholesterol concentration, and blood pressure were observed in vegetarians, having a cardioprotective effect [5]. The risk of death caused by ischemic heart disease is 24% lower in vegetarians than in people who regularly consume meat. The conducted studies suggest that the vegetarian diet—eliminating red meat and poultry—and the lacto-vegetarian diet decrease the level of total cholesterol and low-density lipoprotein (LDL) cholesterol by about 10–15%, whereas the vegan diet—by about 15–25% [6]. Frequently occurring caloric restrictions in the use of vegetarian diets have an effect on life extension and protection against cancer [7]. Likewise, eliminating fried, smoked, or grilled foods, red meat, aflatoxin-contaminated foods, preserved salty meals, and alcohol [8]. Moreover, cancer risk is also reduced by introducing a diet rich in plant foods (e.g., vegetables, beans, fruit, and whole grains) and by reducing consumption of animal fat, meat, and fatty dairy products [9].

Vegetables and fruits are an important source of a wide range of bioactive ingredients and compounds, including antioxidants, vitamins (folic acid, carotenoids), glucosinolates, indoles, isothiocyanates, protease inhibitors, lycopene, phenolic compounds, and flavonoids that exhibit anti-cancer properties [10,11].

Studies conducted by the Homo Homini Opinion Institute in 2013 showed that the number of vegetarians in Poland is over a million, including 1.6% of lacto-vegetarians and 1.6% of vegans. A Centre for Public Opinion Research (CBOS) survey indicated that 26% of Poles followed some form of an elimination diet in 2014. The excluded products featured meat (11%), milk (9%), milk products (8%), fish (4%), and chicken eggs (4%) [12].

The aim of our study was to track the changes associated with the quantity of individual ingredients, including nutrients, fatty acids, amino acids, vitamins, and minerals. To achieve this, it was determined that all of the analyzed diets would include the same caloricity in accordance with the daily need of 2000 kcal.

2. Types of Diets Eliminating Products of Animal Origin, Which Were Analyzed in This Study

2.1. Semi-Vegetarian Diet

The elimination of red meat decreases the frequency of occurrence of type 2 diabetes and regulates carbohydrate metabolism [13,14]. Moreover, the risk of large intestine, breast, and prostate cancers was much higher in omnivores than in vegetarians [5]. Vegetarian diets are associated with increased consumption of vegetables and fruit rich in phytochemicals, dietary fiber, and antioxidants. These substances have a positive influence on health, e.g., they protect the organism against free radicals, simultaneously preventing the formation of neoplasms. Plant diets are also characterized by low concentrations of saturated fatty acids (SFA) and by high concentrations of polyunsaturated fatty acids (PUFA), having a positive influence on the lipid profile [15]. Soy often replaces meat in plant diets as it is a rich source of proteins and phytoestrogens, particularly isoflavones, which may play a protective role in reference to the development of breast cancer [16].

2.2. Lacto-Ovo-Vegetarian Diet

The lacto-ovo-vegetarian diet (LOV) is a type of vegetarianism that eliminates meat, meat products, and fish, but allows for the consumption of eggs and dairy products. People following this type of diet avoid products that simultaneously fulfill two criteria—"when this animal was alive, it had eyes and mommy". Eggs in the diet are a necessary source of fatty acids, wholesome protein, choline, selenium, Vitamin A and B_{12} [17]. One of the constituents of yolk is cholesterol, which for many years has been considered the cause of increased risk of cardiovascular diseases (CVD). Therefore, it is recommended for omnivores to consume no more than 3 yolks per week. The most recent studies indicate that the presence of higher numbers of eggs in the weekly menu (more than 3) is not dangerous to health if other sources of cholesterol are reduced. Yolk is a valuable source of fatty acids, lecithin, choline, xanthophylls, immunoglobulin, and vitamins. Differences in the quantitative content of eggs may occur in the case of a different method of nutrition of hens, and as a result of special fodder additives [18].

2.3. Vegan Diet

The most restrictive type of vegetarianism is veganism, which eliminates all products of animal origin. Food products present in vegan menus include: cereal, fruit, vegetables, nuts, mushrooms, legumes, oils, and plant drinks. People choose the vegan diet mainly due to ethics (protection of animal rights), religious beliefs, and personal health. However, when applied for a longer period of time, this type of nutrition without proper balancing and supplementation results in negative health consequences. The main risk is the deficiency of vitamins and certain minerals that are necessary for the proper functioning of the human body [5]. It has been demonstrated that incorrectly balanced vegan diet may be the cause of the development of numerous neurological disorders, such as fear, depression, brainstorm, neuropathy, chronic tiredness, and insomnia [19]. Vegans also suffer from low concentrations and deficiencies of Vitamin B_{12}, resulting from the exclusion of all products of animal origin, which are their only natural source [20]. B_{12} deficiencies may lead to megaloblastic anemia or demyelinating disease [21]. The vegan diet is characterized by high folate content, positively influencing the correct functioning of hematopoietic and nervous systems [15]. One of the factors that influences the development of CVD and neoplasms is obesity. It has been proven that vegans have lower average BMI than omnivores and other vegetarians, which decreases the risk of heart diseases and mortality resulting from ischemic heart disease [22]. The main sources of proteins in vegans include legumes, which include fiber and phytochemicals that help control glycemia, reducing the risk of developing type 2 diabetes [23].

2.4. Milk-Free Diet

The dairy-free or lactose-free diet is most often applied due to cow milk protein allergy (CMPA), lactose intolerance, lack of availability, and the increasingly popular trend. Products that are eliminated in the menu include milk and dairy products (cheese, yoghurt, cottage cheese, milk kefir, cream, buttermilk). Study results show that only 25% of the world's population can breakdown lactose both in childhood and as adults. In most infants, the activity of intestinal lactase is the strongest in the perinatal period. However, after 2 years (from 2 to 12 years of age), two different groups appear, i.e., people with low and people with high lactase activity. Another case includes people that have the lactose breakdown ability throughout their entire life [24]. There are people that experience lactose intolerance, which is why after the consumption of milk and milk products they suffer from annoying symptoms referring to the gastrointestinal tract, as well as systemic symptoms. Colon microflora ferments the undigested lactose in the gut, leading to the creation of short-chain fatty acids (SCFA), hydrogen, carbon dioxide, and methane. Lactose intolerance symptoms can be untypical, e.g., headaches and dizziness, mouth sores, sore throat caused by the increase in the size of lymph nodes and muscles [25].

Dairy products play a key role in human diet as they are a rich source of vitamins and minerals, especially riboflavin, calcium, and wholesome protein [26,27]. However, dairy is characterized by low content of iron and folic acid [28]. It is recommended to consume lean dairy products due to the reduced content of saturated fatty acids (SFA), which are responsible for the increase of LDL cholesterol, and the increased risk of CVD. It has also been observed that the milk-free diet is helpful in the treatment of intestinal inflammations or acne inversa, which is associated with insulin metabolism [29].

2.5. Fish-Free Diet

Another type of the studied diet is one that eliminates fish. The consumption of fish and seafood has a positive influence on health—a fact that is supported by numerous studies. The main advantage of fish is the content of omega-3 fatty acids, Vitamin D and minerals, including iodine [30]. Some of the most important elements of the diet are fatty sea fish, i.e., salmon, herring, mackerel—the ones that are the richest in eicosapentaenoic acid (EPA) and docosahexaenoic acid (DHA) fatty acids. They influence such factors as the proper functioning of the circulatory system, and they are important for the proper development of the fetus, including the function of its neurons and retina, as well as for the immunity of the entire organism. EPA and DHA fatty acids are also associated with promising results in terms of the prevention and management of body mass and cognitive functions in people with very mild Alzheimer's disease, a better prognosis after ischemic stroke in the treatment of illnesses with an inflammatory basis [31]. Vitamin D_3 is naturally present in many products, but its richest sources include some fatty fish, oil extracted from fish liver, and caviar.

Numerous studies conducted in recent years have demonstrated that the increasing pollution of the environment has an influence on fish and shellfish. The increase in toxic substances originating from farms and the presence of heavy metals in the environment of water animals negatively influence human health via the food chain [32]. Fish feature increased concentrations of metals, such as arsenic, cadmium, chromium, mercury, and iron, which makes people avoid their consumption [33]. Other reasons for the limitation of the consumption of fish include fish and shellfish protein allergies, the specific smell which is unacceptable by some people, as well as high price, particularly of wild fish.

2.6. Regular Diet of Poles

In most cases, the regular diet of Poles (RD) includes all types of animal products, including red meat, white meat, eggs, milk, dairy, and—occasionally—fish. Recent studies have shown that the diet of Poles features excessive total fat content, including SFA and cholesterol. Insufficient consumption of PUFA and dietary fiber has also been observed [34]. Excessive amounts of salt and saccharose are frequently observed. All of these factors may be the cause of the increased risk of CVD the metabolic syndrome progressing together with diabetes and fatty liver. Other studies indicated that the menus of Poles are deficient in terms of folates, Vitamin D, thiamin, and niacin, and when it comes to minerals, they lack the proper concentrations of calcium, potassium, magnesium and iron in the group of women [35,36].

To conclude, our goal was to compare diets in terms of nutrient content, to determine any potential risk to health in the case of diets properly prepared by qualified dieticians, and to establish whether correctly balanced vegetarian diets could still pose a higher risk of nutritional deficiencies than the regular nutritional habits of Poles.

3. Materials and Methods

The study was conducted after receiving the permission of the Bioethics Committee of the Pomeranian Medical University (regulation no. KB-0012/116/15). The subjects gave written informed consent, and their confidentiality and anonymity were protected. The study population consisted of 30 people, including 18 men and 12 women. From this group, menus for the RD diet were selected from 10 people (equal number of sexes), corresponding to the assumed calorific value (2000 kcal). All subjects were Caucasian, potentially healthy with an average age of 39.26 ± 7.86 for women and 40.56 ± 6.43 for men.

3.1. Preparation of Menus

The study features the quantitative analysis of a total of 70 menus, 10 for each type of diet eliminating products of animal origin, as well as RD without elimination. Every menu has been prepared by qualified dieticians with the use of methods avoiding nutritional deficiencies, however, do not include supplementation. The caloricity of the developed menus was ±2000 kcal. The total energy from fat was 25–35%, proteins 10–17%, and carbohydrates complemented the rest of the caloricity. The supply of ingredients with water was not included in any of the menus.

Additionally, 30 menus were collected on the basis of a 24 h interview on nutrition conducted with adults, and the data was incorporated into a dietary program in order to evaluate caloricity and the content of individual nutrients. Out of the menus, those that fulfilled the inclusion criteria of ±2000 kcal were chosen. Then 29 ingredients were analyzed: energy, total protein, plant protein, animal protein, carbohydrates, dietary fiber, fat, cholesterol, omega-3 and omega-6 fatty acids, minerals: Na, K, Ca, P, Mg, Fe, Zn, Cu, Mn; vitamins soluble in water: C, B1, B2, B3, B6, B12; folic acid, vitamins soluble in fats: A, D, E; the participation of energy, fats and carbohydrates. The analysis also includes the concentration of exogenous and relatively exogenous amino acids.

3.2. Quantitative Analysis

The quantitative analysis of the menus was conducted using the Dieta 6 days dietary program, which is recommended by the National Institute of Food and Nutrition. The total caloricity of the menus has been divided into 3 main meals (breakfast, dinner, supper) and 2 additional meals (second breakfast and an evening meal fulfilling a complimentary role).

Types of the analyzed diets, prepared by dieticians:

- The basic diet (BD), which includes all product groups: meat, dairy, eggs, and fish;
- The modification of a vegetarian diet with fish (pescoveget-PV), which eliminates red and white meat, but allows for the consumption of eggs, dairy and fish;
- The milk-free diet (MFD), which eliminates all dairy products;
- The vegan diet (VEGAN), which eliminates all products of animal origin: meat, eggs, dairy, fish and honey;
- Diet that eliminates fish, seafood and shellfish (FFD);
- The lacto-ovo-vegetarian diet (LOV), which eliminates meat and fish, allowing for the consumption of eggs and dairy;
- The regular diet (RD); data collected on the basis of a 24 h nutrition interview of 10 Polish adults, with the selection of menus of 2000 kcal caloricity;

After creating a basic diet (BD) containing all protein sources from the diet, subsequent products were eliminated and replaced with substitutes. They were: lean dairy products (yoghurt, mozzarella, cottage cheese) and/or eggs and/or fish: cod, mackerel, salmon, tuna, herring, trout and/or vegetable milk, and legumes, tofu, seeds: almonds, nuts, cocoa, sesame, poppy seed, linseed, amaranth, pumpkin, sunflower seeds. The appropriate caloric value and the percentage of protein, fat, and carbohydrates in the diet were respected. The diets have been planned in order to be similar to the Polish diet with the best possible use of products available on the market that complemented possible shortages. The average content of nutrients was compared to the updated nutritional norms for the Polish population, developed by the National Institute of Food and Nutrition in 2017 [37]. Table 1 presents the current norms for nutrients developed by the Institute and determined at the level of the estimated average requirement (EAR). The table features ingredients that are present in Dieta 6d software.

Table 1. Nutrient norms for men and women over 19 years of age, determined on the basis of the EAR.

Nutrients	Women	Men
Vitamin A (µg)	500	630
Vitamin D (µg)	15	15
Vitamin E (mg)	8	10
Vitamin C (mg)	60	75
Vitamin B_1 (thiamin) (mg)	0.9	1.1
Vitamin B_2 (riboflavin) (mg)	0.9	1.1
Vitamin B_3 (niacin) (mg)	11	12
Vitamin B_6 (mg)	1.1	1.1
Folates (µg)	320	320
Vitamin B_{12} (cobalamin) (µg)	2.0	2.0
Calcium (mg)	800	800
Phosphorus (mg)	580	580
Magnesium (mg)	255	330
Iron (mg)	8	6
Zinc (mg)	6.8	9.4
Copper (mg)	0.7	0.7
Iodine (µg)	95	95
Manganese (mg)	1.8	2.3
Sodium (mg)	1500	1500
Potassium (mg)	3500	3500

EAR, estimated average requirement.

3.3. Statistical Analysis

The statistical analysis was conducted using STATISTICA 13.3 (StatSoft, Cracow, Poland). The average ($\overline{x}$) and standard deviation (SD) were calculated for all individual nutrients in the analyzed diets. Analysis of variance (ANOVA) was used to determine the differences in the content of individual nutrients between the diets. Statistical significance was established at $p \leq 0.05$. In accordance with the assumptions of the study, we observed no statistically significant differences in the caloricity of the analyzed diet types ($p \geq 0.05$).

4. Results

4.1. Analysis of the Content of Proteins and Amino Acids

Statistically significant differences were observed in terms of plant and animal protein between MFD and the other diets (Figure 1). In the case of FFD, the content of animal protein in the menu significantly differs from RD, VEGAN, and PV, whereas for plant protein, significant differences are observed with reference to RD and VEGAN. The basic diet contains statistically significantly more animal protein than LOV, VEGAN, PV, and MFD ($p < 0.05$), but significantly lower content of plant protein than VEGAN and MFD. The content of plant protein in the regular diet is significantly lower than in the case of other diets. It has also been observed that there are statistically significant differences in the content of plant protein in the following pairs: LOV and VEGAN; VEGAN and PV (Table 2).

Table 2. The average content of proteins between the diets (g/day).

	Total Protein						
	MFD	**FFD**	**BD**	**RD**	**LOV**	**VEGAN**	**PV**
$\overline{x} \pm$ SD (g/day)	76.3 ± 4.13	79.2 ± 6.18	81.5 ± 4.38	77 ± 10.04	75.5 ± 5.76	66.1 ± 4.52	73.2 ± 7.94
MFD	-	0.3196	0.0770	0.8023	0.7777	**0.0008**	0.2904
FFD	0.3196	-	0.4298	0.4550	0.2029	**<0.0001**	0.0426
BD	0.0770	0.4298	-	0.1270	**0.0415**	**<0.0001**	**0.0057**
RD	0.8023	0.4550	0.1270	-	0.5946	**0.0003**	0.1924
LOV	0.7777	0.2029	**0.0415**	0.5946	-	**0.0018**	0.4368
VEGAN	**0.0008**	**<0.0001**	**<0.0001**	**0.0003**	**0.0018**	-	**0.0162**
PV	0.2904	0.0426	**0.0057**	0.1924	0.4368	**0.0162**	-

$\overline{x}$—average value; SD—standard deviation; bold—statistically significant differences ($p \leq 0.05$); MFD, milk-free diet; FFD, diet that eliminates fish, seafood and shellfish; BD, basic diet; RD, regular diet; LOV, lacto-ovo-vegetarian diet; PV, pascoveget.

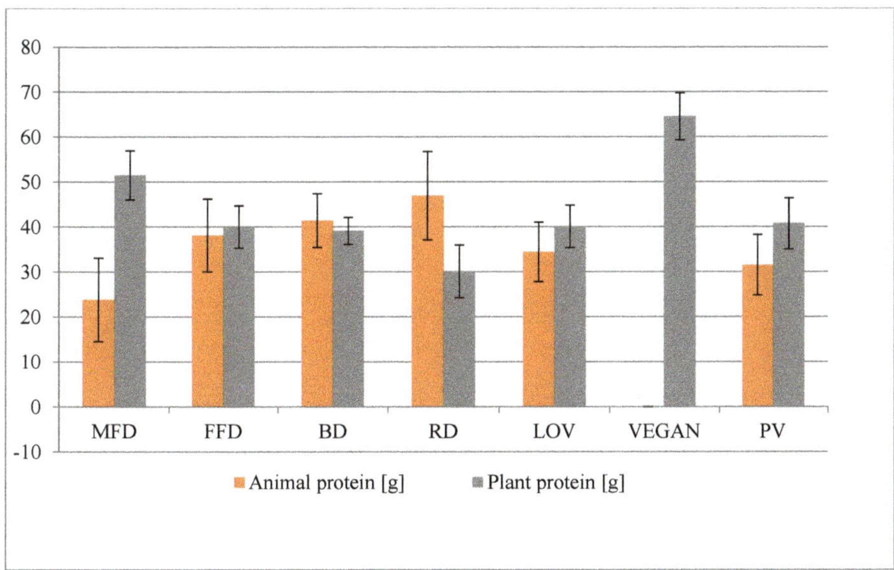

Figure 1. The average content of total plant and animal protein in the diets.

The lowest content of all exogenous amino acids was observed in the VEGAN diet (Table 3). Statistically significant differences were observed in the content of isoleucine, leucine, lysine, methioninee, phenylalanine, threonine, tryptophan, and valine between VEGAN and other diets. In terms of isoleucine, PV differed significantly from FFD and BD. MFD and BD showed a difference in the content of isoleucine, leucine, lysine, methionine, phenylalanine, and valine. In terms of leucine, significant differences were observed between BD and PV, and between MFD and LOV. FFD and MFD statistically significantly differed in the quantity of leucine, lysine, phenylalanine and valine. Statistical significance in terms of lysine content was observed for BD and RD, BD and LOV, BD and PV, FFD and PV. In terms of methioninee—MFD and RD, BD and PV, RD and PV. For threonine—BD and LOV, RD and LOV, PV and FFD, BD and RD. The content of tryptophan significantly differed between PV and FFD, BD. PV and BD were different in terms of valine content (Table 4).

Table 3. The average content of exogenous amino acids in the analyzed diets (mg/day).

Amino Acid (mg)	MFD $\bar{x}$ ± SD	FFD $\bar{x}$ ± SD	BD $\bar{x}$ ± SD	RD $\bar{x}$ ± SD	LOV $\bar{x}$ ± SD	VEGAN $\bar{x}$ ± SD	PV $\bar{x}$ ± SD
Isoleucine	3413 ± 375	3703 ± 444	3829 ± 352	3654 ± 492	3528 ± 282	2737 ± 210	3360 ± 410
Leucine	5428 ± 536	6145 ± 621	6298 ± 553	5815 ± 843	5970 ± 533	4570 ± 407	5748 ± 662
Lysine	4180 ± 557	5008 ± 735	5317 ± 506	4710 ± 740	4518 ± 486	3121 ± 442	4397 ± 646
Methionine	1543 ± 298	1721 ± 249	1822 ± 231	1815 ± 256	1642 ± 197	940 ± 88.3	1530 ± 213
Phenyl-alanine	2962 ± 250	3602 ± 314	3649 ± 293	3520 ± 502	3558 ± 291	2686 ± 301	3421 ± 355
Threonine	2931 ± 85	3043 ± 380	3198 ± 302	2147 ± 434	2852 ± 248	2376 ± 220	2739 ± 328
Tryptophan	972 ± 118	1030 ± 118	1050 ± 107	965 ± 142	957 ± 91.3	810 ± 69	926 ± 126
Valine	4020 ± 426	4532 ± 459	4640 ± 427	4263 ± 585	4396 ± 366	3267 ± 242	4234 ± 477

$\bar{x}$—average value; SD—standard deviation; bold—statistically significant differences ($p \leq 0.05$).

Table 4. The differences in the content of exogenous amino acids.

Isoleucine	MFD	FFD	BD	RD	LOV	VEGAN	PV
MFD	-	0.0906	**0.0164**	0.1585	0.4969	**0.0002**	0.7561
FFD	0.0906	-	0.4583	0.7714	0.3044	**<0.0001**	**0.0465**
BD	**0.0164**	0.4583	-	0.3033	0.0796	**<0.0001**	**0.0072**
RD	0.1585	0.7714	0.3033	-	0.4599	**<0.0001**	0.0869
LOV	0.4969	0.3044	0.0796	0.4599	-	**<0.0001**	0.3234
VEGAN	**0.0002**	**<0.0001**	**<0.0001**	**<0.0001**	**<0.0001**	-	**0.0005**
PV	0.7561	**0.0465**	**0.0072**	0.0869	0.3234	**0.0005**	-

LEUCINE	MFD	FFD	BD	RD	LOV	VEGAN	PV
MFD	-	**0.0104**	**0.0021**	0.1587	**0.0500**	**0.0024**	0.2434
FFD	**0.0104**	-	0.5744	0.2285	0.5219	**<0.0001**	0.1479
BD	**0.0021**	0.5744	-	0.0798	0.2314	**<0.0001**	**0.0466**
RD	0.1587	0.2285	0.0798	-	0.5693	**<0.0001**	0.8044
LOV	**0.0500**	0.5219	0.2314	0.5693	-	**<0.0001**	0.4148
VEGAN	**0.0024**	**<0.0001**	**<0.0001**	**<0.0001**	**<0.0001**	-	**<0.0001**
PV	0.2434	0.1479	**0.0466**	0.8044	0.4148	**<0.0001**	-

LYSINE	MFD	FFD	BD	RD	LOV	VEGAN	PV
MFD	-	**0.0029**	**<0.0001**	0.0517	0.2111	**0.0002**	0.4193
FFD	**0.0029**	-	0.2513	0.2708	0.0718	**<0.0001**	**0.0259**
BD	**<0.0001**	0.2513	-	**0.0267**	**0.0040**	**<0.0001**	**0.0010**
RD	0.0517	0.2708	**0.0267**	-	0.4141	**<0.0001**	0.2462
LOV	0.2111	0.0718	**0.0040**	0.4141	-	**<0.0001**	0.6540
VEGAN	**0.0002**	**<0.0001**	**<0.0001**	**<0.0001**	**<0.0001**	-	**<0.0001**
PV	0.4193	**0.0259**	**0.0010**	0.2462	0.6540	**<0.0001**	-

METHIONINE	MFD	FFD	BD	RD	LOV	VEGAN	PV
MFD	-	0.0860	**0.0079**	**0.0097**	0.3362	**<0.0001**	0.8967
FFD	0.0860	-	0.3222	0.3604	0.4411	**<0.0001**	0.0655
BD	**0.0079**	0.3222	-	0.9393	0.0810	**<0.0001**	**0.0055**
RD	**0.0097**	0.3604	0.9393	-	0.0947	**<0.0001**	**0.0068**
LOV	0.3362	0.4411	0.0810	0.0947	-	**<0.0001**	0.2758
VEGAN	**<0.0001**	**<0.0001**	**<0.0001**	**<0.0001**	**<0.0001**	-	**<0.0001**
PV	0.8967	0.0655	**0.0055**	**0.0068**	0.2758	**<0.0001**	-

PHENYLALANINE	MFD	FFD	BD	RD	LOV	VEGAN	PV
MFD	-	**<0.0001**	**<0.0001**	**0.0005**	**0.0002**	0.0729	**0.0035**
FFD	**<0.0001**	-	0.7588	0.5861	0.7736	**<0.0001**	0.2358
BD	**<0.0001**	0.7588	-	0.3954	0.5524	**<0.0001**	0.1372
RD	**0.0005**	0.5861	0.3954	-	0.7970	**<0.0001**	0.5183
LOV	**0.0002**	0.7736	0.5524	0.7970	-	**<0.0001**	0.3673
VEGAN	0.0729	**<0.0001**	**<0.0001**	**<0.0001**	**<0.0001**	-	**<0.0001**
PV	**0.0035**	0.2358	0.1372	0.5183	0.3673	**<0.0001**	-

THREONINE	MFD	FFD	BD	RD	LOV	VEGAN	PV
MFD	-	0.4410	0.0679	0.1386	0.5852	**0.0003**	0.1860
FFD	0.4410	-	0.2833	0.4715	0.1902	**<0.0001**	**0.0386**
BD	0.0679	0.2833	-	0.7217	**0.0191**	**<0.0001**	**0.0022**
RD	0.1386	0.4715	0.7217	-	**0.0447**	**<0.0001**	**0.0061**
LOV	0.5852	0.1902	**0.0191**	**0.0447**	-	**0.0015**	0.4334
VEGAN	**0.0003**	**<0.0001**	**<0.0001**	**<0.0001**	**0.0015**	-	**0.0140**
PV	0.1860	**0.0386**	**0.0022**	**0.0061**	0.4334	**0.0140**	-

TRYPTOPHAN	MFD	FFD	BD	RD	LOV	VEGAN	PV
MFD	-	0.2565	0.1301	0.8809	0.7632	**0.0020**	0.3587
FFD	0.2565	-	0.6988	0.1998	0.1527	**<0.0001**	**0.0426**
BD	0.1301	0.6988	-	0.0971	0.0710	**<0.0001**	**0.0167**
RD	0.8809	0.1998	0.0971	-	0.8796	**0.0031**	0.4417
LOV	0.7632	0.1527	0.0710	0.8796	-	**0.0048**	0.5362
VEGAN	**0.0020**	**<0.0001**	**<0.0001**	**0.0031**	**0.0048**	-	**0.0245**
PV	0.3587	**0.0426**	**0.0167**	0.4417	0.5362	**0.0245**	-

VALINE	MFD	FFD	BD	RD	LOV	VEGAN	PV
MFD	-	**0.0110**	**0.0023**	0.2188	0.0591	**0.0003**	0.2782
FFD	**0.0110**	-	0.5824	0.1733	0.4882	**<0.0001**	0.1321
BD	**0.0023**	0.5824	-	0.0581	0.2159	**<0.0001**	**0.0418**
RD	0.2188	0.1733	0.0581	-	0.4990	**<0.0001**	0.8825
LOV	0.0591	0.4882	0.2159	0.4990	-	**<0.0001**	0.4105
VEGAN	**0.0003**	**<0.0001**	**<0.0001**	**<0.0001**	**<0.0001**	-	**<0.0001**
PV	0.2782	0.1321	**0.0418**	0.8825	0.4105	**<0.0001**	-

$\bar{x}$—average value; bold—statistically significant differences ($p \leq 0.05$).

4.2. Analysis of Fat Content

The highest content of total fats, saturated fats, and monounsaturated fats was observed in the regular diet (RD). The VEGAN diet was characterized by the highest average content of PUFA (Figure 2). Statistically significant differences were observed in terms of fat content between RD and the other diets, between LOV and MFD, and between LOV and VEGAN (Figure 2, Table 5).

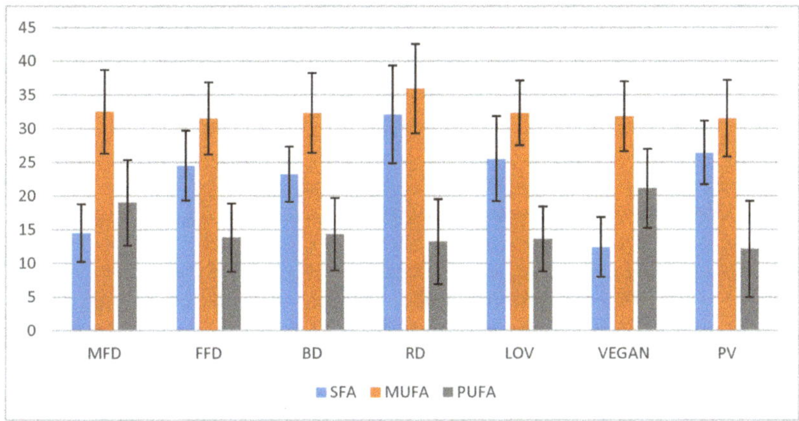

Figure 2. The content of saturated fatty acids (SFA), monounsaturated fatty acids (MUFA), and polyunsaturated fatty acids (PUFA) in diets (g/day).

Table 5. The comparison of the average content of total fats, PUFA, and cholesterol between the diets (g/day; mg/day).

	MFD	FFD	BD	RD	LOV	VEGAN	PV
Total Fats (g/day)							
$\bar{x} \pm SD$	71.7 ± 3.5	75.1 ± 3.0	74.8 ± 3.1	86.8 ± 8.6	76.9 ± 4.1	70.7 ± 7.1	75.1 ± 2.6
MFD	-	0.1461	0.1723	**<0.0001**	**0.0248**	0.6478	0.1397
FFD	0.1461	-	0.9278	**<0.0001**	0.4108	0.0580	0.9809
BD	0.1723	0.9278	-	**<0.0001**	0.3616	0.0705	0.9087
RD	**<0.0001**	**<0.0001**	**<0.0001**	-	**<0.0001**	**<0.0001**	**<0.0001**
LOV	**0.0248**	0.4108	0.3616	**<0.0001**	-	0.076	0.4245
VEGAN	0.6478	0.0580	0.0705	**<0.0001**	0.076	-	0.0551
PV	0.1397	0.9809	0.9087	**<0.0001**	0.4245	0.0551	-
Long-Chain Pufa Acids (g/day)							
$\bar{x} \pm SD$	0.601 ± 1.1	0.032 ± 0.03	0.602 ± 1.1	0.106 ± 0.16	0.052 ± 0.04	0.004 ± 0.01	0.024 ± 0.03
MFD	-	**0.0368**	0.9977	0.0682	**0.0438**	**0.0289**	**0.0344**
FFD	**0.0368**	-	**0.0365**	0.7821	0.9397	0.9184	0.9772
BD	0.9977	**0.0365**	-	0.0678	**0.0435**	**0.0287**	**0.0342**
RD	0.0682	0.7821	0.0678	-	0.8407	0.7047	0.7602
LOV	**0.0438**	0.9397	**0.0435**	0.8407	-	0.8586	0.9169
VEGAN	**0.0289**	0.9184	**0.0287**	0.7047	0.8586	-	0.9411
PV	**0.0344**	0.9772	**0.0342**	0.7602	0.9169	0.9411	-
Cholesterol (mg/day)							
$\bar{x} \pm SD$	159.7 ± 98.8	202.4 ± 105.4	217.4 ± 101.3	369.9 ± 185	306.5 ± 150.9	0.1 ± 0.0	190.9 ± 95.2
MFD	-	0.4215	0.2777	**0.0002**	**0.0070**	**0.0036**	0.5562
FFD	0.4215	-	0.7760	**0.0036**	0.0525	**0.0003**	0.8285
BD	0.2777	0.7760	-	**0.0052**	0.0958	**<0.0001**	0.6165
RD	**0.0002**	**0.0036**	**0.0052**	-	0.2335	**<0.0001**	**0.0012**
LOV	**0.0070**	0.0525	0.0958	0.2335	-	**<0.0001**	**0.0319**
VEGAN	**0.0036**	**0.0003**	**<0.0001**	**<0.0001**	**<0.0001**	-	**0.0006**
PV	0.5562	0.8285	0.6165	**0.0012**	**0.0319**	**0.0006**	-

$\bar{x}$—average value; SD—standard deviation; bold—statistically significant differences ($p \leq 0.05$).

Statistically significant differences were observed in the content of long-chain PUFA between the following diets: FFD and MFD, FFD and BD, LOV and MFD, LOV and BD, VEGAN and MFD, VEGAN and BD, PV and MFD, and PV and BD (Table 5). The lowest content of long-chain PUFA was observed in VEGAN (Figure 2).

The average content of cholesterol was the highest in RD and exceeded the recommended daily intake, whereas VEGAN featured very little amounts of this constituent (Table 5). Statistically significant differences in the content of this constituent were observed between VEGAN and all other diets. In terms of cholesterol content, RD differed from MFD, FFD, BD, VEGAN, and PV; LOV differed in comparison to MFD, VEGAN, and PV, whereas PV was different when compared to RD, LOV, and VEGAN (Table 5).

4.3. Analysis of the Content of Carbohydrates

The highest content of carbohydrates (302.7 ± 22.17 g) was observed in VEGAN, the lowest (257.9 ± 24.29 g) in the RD. RD turned out to be deficient in terms of dietary fiber while other diets contained an average of over 35 g/day of this constituent: MFD—41 g, FFD—37 g, BD—37 g, LOV—37 g, and PV 38 g. Statistically significant differences were observed between RD and the other diets. VEGAN differed in terms of fiber content when compared to FFD, BD, RD, and LOV.

Statistically significant differences were observed in the content of saccharose between RD and the other diets. RD was characterized by the highest average content of this constituent and was at the level of 40 ± 13.73 g.

The lowest content of lactose was observed in MFD and VEGAN. No statistically significant differences were observed in the average content of starch in the analyzed diets (Table 6).

Table 6. The comparison of the average content of carbohydrates, dietary fiber, saccharose, lactose, and starch between the diets (g/day).

Total Carbohydrates (g/day)							
	MFD	**FFD**	**BD**	**RD**	**LOV**	**VEGAN**	**PV**
$\bar{x} \pm SD$	293.2 ± 16.54	290.6 ± 7.01	288.8 ± 7.58	257.9 ± 24.29	284.5 ± 13.19	304.7 ± 22.17	293 ± 9.04
MFD	-	0.7082	0.5285	<0.0001	0.2160	0.1059	0.9780
FFD	0.7082	-	0.7974	<0.0001	0.3855	**0.0480**	0.7287
BD	0.5285	0.7974	-	<0.0001	0.5401	**0.0264**	0.5466
RD	**<0.0001**	**<0.0001**	**<0.0001**	-	0.0003	**<0.0001**	**<0.0001**
LOV	0.2160	0.3855	0.5401	0.0003	-	**0.0053**	0.2262
VEGAN	0.1059	**0.0480**	**0.0264**	**<0.0001**	0.0053	-	0.1003
PV	0.9780	0.7287	0.5466	**<0.0001**	0.2262	0.1003	-

Dietary Fiber (g/day)							
	MFD	**FFD**	**BD**	**RD**	**LOV**	**VEGAN**	**PV**
$\bar{x} \pm SD$	41 ± 5.04	38 ± 5.05	38 ± 4.9	17 ± 3.97	37 ± 5.66	46 ± 7.77	38 ± 4.78
MFD	-	0.1621	0.1596	**<0.0001**	0.1147	0.0677	0.2029
FFD	0.1621	-	0.9934	**<0.0001**	0.8540	**0.0017**	0.8985
BD	0.1596	0.9934	-	**<0.0001**	0.8606	**0.0017**	0.8919
RD	**<0.0001**	**<0.0001**	**<0.0001**	-	**<0.0001**	**<0.0001**	**<0.0001**
LOV	0.1147	0.8540	0.8606	**<0.0001**	-	**0.0010**	0.7554
VEGAN	0.0677	**0.0017**	**0.0017**	**<0.0001**	0.0010	-	**0.0025**
PV	0.2029	0.8985	0.8919	**<0.0001**	0.7554	**0.0025**	-

Saccharose (g/day)							
	MFD	**FFD**	**BD**	**RD**	**LOV**	**VEGAN**	**PV**
$\bar{x} \pm SD$	22.97 ± 7.59	24.7 ± 7.69	24.7 ± 7.74	40 ± 13.73	22.86 ± 7.92	23.59 ± 8.11	24.5 ± 7.6
MFD	-	0.6637	0.6653	**<0.0001**	0.9780	0.8771	0.7016
FFD	0.6637	-	0.9982	**0.0003**	0.6438	0.7791	0.9587
BD	0.6653	0.9982	-	**0.0003**	0.6454	0.7808	0.9605
RD	**<0.0001**	**0.0003**	**0.0003**	-	**<0.0001**	**0.0001**	**0.0002**
LOV	0.9780	0.6438	0.6454	**<0.0001**	-	0.8554	0.6813
VEGAN	0.8771	0.7791	0.7808	**0.0001**	0.8554	-	0.8191
PV	0.7016	0.9587	0.9605	**0.0002**	0.6813	0.8191	-

Table 6. *Cont.*

			Total Carbohydrates (g/day)				
			Lactose (g/day)				
	MFD	**FFD**	**BD**	**RD**	**LOV**	**VEGAN**	**PV**
$\bar{x} \pm$ SD	0.051 ± 0.088	12.4 ± 7.42	11.87 ± 7.27	4.41 ± 4.56	12.07 ± 6.72	0.049 ± 0.09	12.61 ± 7.48
MFD	-	**<0.0001**	**<0.0001**	0.0940	**<0.0001**	0.9994	**<0.0001**
FFD	**<0.0001**	-	0.8368	**0.0028**	0.9005	**<0.0001**	0.9349
BD	**<0.0001**	0.8368	-	**0.0050**	0.9355	**<0.0001**	0.7737
RD	0.0940	**0.0028**	**0.0050**	-	**0.0040**	0.0939	**0.0022**
LOV	**<0.0001**	0.9005	0.9355	**0.0040**	-	**<0.0001**	0.8363
VEGAN	0.9994	**<0.0001**	**<0.0001**	0.0939	**<0.0001**	-	**<0.0001**
PV	**<0.0001**	0.9349	0.7737	**0.0022**	0.8363	**<0.0001**	-
			Starch (g/day)				
	MFD	**FFD**	**BD**	**RD**	**LOV**	**VEGAN**	**PV**
$\bar{x} \pm$ SD	157.7 ± 27.8	147.8 ± 31.3	147.0 ± 29.3	166.9 ± 17.4	145.6 ± 30.3	161.3 ± 20.3	149.6 ± 33.3
MFD	-	0.4250	0.3896	0.4573	0.3329	0.7714	0.5149
FFD	0.4250	-	0.9496	0.1259	0.8633	0.2778	0.8828
BD	0.3896	0.9496	-	0.1115	0.9132	0.2512	0.8333
RD	0.4573	0.1259	0.1115	-	0.0897	0.6499	0.1656
LOV	0.3329	0.8633	0.9132	0.0897	-	0.2096	0.7494
VEGAN	0.7714	0.2778	0.2512	0.6499	0.2096	-	0.3474
PV	0.5149	0.8828	0.8333	0.1656	0.7494	0.3474	-

$\bar{x}$—average value; SD—standard deviation; bold—statistically significant differences ($p \leq 0.05$).

4.4. Analysis of Minerals

RD included the highest average content of sodium, which was a few times higher than the recommended dose. The lowest content of sodium was observed in VEGAN. Statistically significant differences were observed between RD and the other diets. In terms of sodium content, VEGAN differed from FFD, BD, RD, LOV, and PV (Table 7).

Table 7. The average content of minerals in the analyzed diets.

Mineral	MFD $\bar{x} \pm$ SD	FFD $\bar{x} \pm$ SD	BD $\bar{x} \pm$ SD	RD $\bar{x} \pm$ SD	LOV $\bar{x} \pm$ SD	VEGAN $\bar{x} \pm$ SD	PV $\bar{x} \pm$ SD
Sodium (mg)	1175 ± 519.1	1501 ± 575.1	1414 ± 496.1	3527 ± 813	1499 ± 408.4	915 ± 290.5	1490 ± 481
Potassium (mg)	4324.2 ± 743	4242.3 ± 583	4390 ± 616	2594.7 ± 594	4037.5 ± 63	4378.4 ± 845.5	4052 ± 473
Calcium (mg)	452.4 ± 73.8	1281.7 ± 239.3	1169.1 ± 310.2	444.1 ± 157.7	1363.1 ± 267	610.2 ± 107.4	1360.7 ± 299.2
Phosphorus (mg)	1509.5 ± 121.4	1704.8 ± 184.5	1698 ± 189.7	1099.6 ± 178.9	1674.4 ± 118.1	1449.7 ± 145.2	1674.4 ± 193.7
Magnesium (mg)	558.6 ± 66.1	505.3 ± 60.7	508.7 ± 65.2	261.1 ± 70.2	484.8 ± 71.6	589.3 ± 49.1	498.5 ± 72.2
Iron (mg)	18.2 ± 2.2	15.3 ± 1.5	15.4 ± 1.3	11.2 ± 2.2	15.7 ± 1.5	19.8 ± 1.9	15.3 ± 2
Zinc (mg)	12.1 ± 1.9	12.9 ± 1.6	12.6 ± 1.5	10.6 ± 1.8	12.9 ± 1.4	12.6 ± 1.2	12.8 ± 1.6
Copper (mg)	2.4 ± 0.3	2.1 ± 0.3	2.1 ± 0.3	1.0 ± 0.2	2.1 ± 04	2.7 ± 0.4	2.2 ± 0.3
Manganese (mg)	8.0 ± 1.8	7.2 ± 1.7	7.0 ± 1.6	3.0 ± 1.1	7.1 ± 1.3	9.0 ± 1.5	7.2 ± 1.6
Iodine (µg)	62.8 ± 46.4	46.9 ± 20.3	57.9 ± 19.1	87.2 ± 45.0	37.6 ± 9.2	29.4 ± 10.3	39.2 ± 11.1

$\bar{x}$—average value; SD—standard deviation.

The regular diet (RD) did not fulfil the recommended dietary allowances (RDA) norm with reference to some minerals, such as potassium, magnesium, and iodine, and their average content was: potassium 2594.7 ± 594 mg, calcium 444.1 ± 157.7 mg, magnesium 261.1 ± 70.2 mg, and iodine 87.2 ± 45.0 µg. Insufficient level of calcium was also observed in MFD and VEGAN. FFD and BD contained incorrect quantities of calcium for pregnant women and children (according to the RDA norm). The highest average content of calcium was observed in LOV (1363.1 mg) The average requirement for phosphorus for the group (according to EAR) was fulfilled by all diets, but the lowest value was observed in RD (1099.6 mg), and the highest in FFD (1704.8 mg). The average content of phosphorus in RD does not fulfill the recommended daily norm for children, pregnant, and breastfeeding women (under 19 years of age). Magnesium content in RD did not meet the nutritional norm for men. The highest average content of iron was observed in VEGAN (19.8 mg), whereas the lowest in RD (11.2 mg). None of the analyzed diets fulfilled the recommended norm for iron for pregnant women. Women aged <50 should consume 18 mg of iron every day. Lower levels of this element were observed in FFD, BD, RD, LOV, and PV. RD also did not fulfill the norm for iron for

13–18 year-olds—boys and menstruating girls. The content of zinc was comparable in the analyzed diets, but the lowest content was observed in RD—10.6 mg. All diets featured lower than recommended content of zinc for breastfeeding women under 19. RD did not fulfill RDA for men, pregnant women, and breastfeeding women. The highest content of copper and manganese was observed in VEGAN (2.7 mg and 9.0 mg, respectively), while the lowest in RD (1.0 mg and 3.0 mg, respectively). RD featured lower than recommended content of copper for lactating women. All of the analyzed diets featured insufficient quantities of iodine when compared to nutritional norms (S1).

4.5. Vitamin Content Analysis

The lowest content of retinol was observed in VEGAN (22.7 µg), the highest in RD (682.7 µg). The most beta-carotene rich diet was BD (8942.3 µg), whereas RD included the lowest amounts of this constituent (1917.6 µg). RD did not meet the nutritional norm for Vitamin E in men, pregnant women and breastfeeding women. All of the analyzed diets fulfilled the norm for thiamin (Vitamin B_1). RD and VEGAN featured insufficient amounts of riboflavin (Vitamin B_2), which covers the needs of lactating women. MFD was the only diet that included the appropriate content of niacin (Vitamin B_3). Insufficient average amount of this vitamin for pregnant women was observed in FFD, BD, RD, LOV, VEGAN, and PV, whereas in the case of breastfeeding women, deficiencies were observed in BD, RD, LOV, VEGAN, and PV. RD did not fulfill RDA in terms of Vitamin B_6 content for pregnant women and lactating women. The lowest concentration of Vitamin C was observed in RD—70.9 mg. The recommended daily consumption of folates for pregnant women is 600 µg. All of the analyzed diets had lower values of this constituent. In the case of breastfeeding women, the recommended amount is 500 µg, but the norm was not fulfilled by MFD, FFD, BD, RD, and PV. VEGAN featured the lowest content of Vitamin B_{12} (2.2 µg), and this amount is not enough to cover the recommended daily consumption for men and women, including those that are pregnant or breastfeeding. FFD and PV were characterized by the lowest content of cobalamin with reference to lactating women. The highest content of this vitamin was observed in RD and MFD (4.1 µg). None of the analyzed diets fulfilled the norm for Vitamin D (Table 8 and Table S2).

Table 8. Comparison of the average content of vitamins in the analyzed diets.

Vitamin	MFD $\bar{x} \pm$ SD	FFD $\bar{x} \pm$ SD	BD $\bar{x} \pm$ SD	RD $\bar{x} \pm$ SD	LOV $\bar{x} \pm$ SD	VEGAN $\bar{x} \pm$ SD	PV $\bar{x} \pm$ SD
Vitamin A (µg)	1573.8 ± 682.9	1697.8 ± 775.6	1727.1 ± 739.7	1000 ± 491.6	1769.5 ± 766.3	1473.6 ± 709.2	1665.4 ± 641.3
Retinol (µg)	126.1 ± 96.3	243.5 ± 87.9	242.1 ± 83.6	682.7 ± 435.5	352.7 ± 131.3	22.7 ± 31.9	286.8 ± 68.8
β-carotene (µg)	8463.2 ± 4300.5	8757.8 ± 4519	8942.3 ± 4514.3	1917.6 ± 2169.6	8533.6 ± 4049.1	8462.3 ± 4297	8283.2 ± 3901
Vitamin E (mg)	19.4 ± 3.8	15.3 ± 3.4	15.7 ± 3.5	9.6 ± 4.8	15.5 ± 3.2	18.7 ± 3.4	15.1 ± 3.5
Thiamin (mg)	1.9 ± 0.3	1.7 ± 0.4	1.7 ± 0.4	1.5 ± 0.4	1.5 ± 0.3	2 ± 0.4	1.6 ± 0.3
Riboflavin (mg)	1.6 ± 0.2	1.9 ± 0.4	1.9 ± 0.4	1.4 ± 0.3	2.0 ± 0.4	1.4 ± 0.3	2 ± 0.3
Niacin (mg)	21.7 ± 5.6	17.6 ± 4.6	10.1 ± 5.2	16 ± 4.4	13.6 ± 3.6	15.6 ± 3.3	13.6 ± 3.0
Vitamin B_6 (mg)	3.0 ± 0.5	2.7 ± 0.5	2.9 ± 0.6	1.7 ± 0.5	2.5 ± 0.4	2.7 ± 0.5	2.4 ± 0.4
Vitamin C (mg)	208.3 ± 52.3	216.8 ± 55.3	219.9 ± 56.9	70.9 ± 47	227.9 ± 69	210.1 ± 53.9	210.6 ± 46.8
Folates (µg)	496.4 ± 61.3	485.4 ± 71.2	490.6 ± 74.2	216.5 ± 57.6	507.9 ± 78.1	520.3 ± 87.8	494.6 ± 65.1
Vitamin B_{12} (µg)	4.1 ± 1.8	2.7 ± 0.8	3.5 ± 1.3	4.1 ± 2.1	3.1 ± 0.9	2.2 ± 1.3	2.6 ± 0.8
Vitamin D (µg)	4.3 ± 4.7	1.0 ± 0.8	3.4 ± 4.3	3.5 ± 2.3	1.5 ± 0.7	1.0 ± 0.8	1.0 ± 0.7

$\bar{x}$—average value, SD—standard deviation.

5. Discussion

The nutritional habits of Poles deviate in many ways from the rules of rational nutrition. The diets are poor in vegetables, fruit, and wholegrain products, which results in the deficiency of some vitamins and minerals. Despite the same caloricity of the diets, there were differences between them in protein content. The source of proteins in a diet include products of animal and plant origin. The vegan (VEGAN) diet contained no products of animal origin and as such, it had the highest content of plant protein. Higher consumption of plant products is associated with the higher provision of individual constituents, such as fiber, potassium, and magnesium. Therefore, the decrease in the prevalence of CVD results from a number of factors, not just the consumption of plant protein [38]. The replacement of 1 standard portion of red meat (85 g) with three different plant sources of protein decreases the risk of coronary heart disease (CHD) by 13–30% according to the Nurses 'Health Study [39], and by

7–19% in the combined control analyses Nurses 'Health Study and Health Professionals Follow-Up [40]. The type of the consumed protein is important due to the various contribution of exogenous amino acids between plant and animal products [41]. Our study confirmed that the vegan diet and its derivatives pose a risk of insufficient supply of endogenous amino acids, relatively exogenous amino acids, as well as purely exogenous amino acids (leucine, isoleucine, lysine, methionine, threonine, phenylalanine, tryptophan, and valine). In the case of relatively exogenous amino acids, which include histidine, arginine, and serine, these amino acids can be produced in the body, but in exceptional situations, such as various illnesses, stressful events, or the period of quick growth, they should be supplied in appropriate amounts together with food [42]. In the context of the vegetarian diet, lysine is an important amino acid that requires special attention. In products of plant origin, its content is limited. To increase the supply of lysine, the vegetarian and (especially) vegan diets should be enriched with nuts and/or soy seeds. At the same time, appropriate supply of lysine has an influence on the decrease of the risk of heart diseases and some neoplasms as a result of the limitation of the activity of enzymes responsible for the lipogenesis and synthesis of cholesterol. Lysine also contributes to the reduction of the concentration of insulin-like growth factors (IGF) [43]. Another amino acid whose decreased supply is observed in the vegetarian diet is methionine. During the course of various transformations, methionine is transformed into taurine and homocysteine. For many people following the vegetarian diet, this amino acid reduces the assimilation of other amino acids [44]. When it comes to mental health, tryptophan is important. This amino acid is necessary for the production of serotonin, which is responsible for feeling well, the regulation of sleep, and it also prevents hyperactivity in children. Moreover, it is transformed into melatonin, and it also influences the secretion of hormones that support the synthesis of pyridoxine and niacin. The richest source of tryptophan is turkey meat, milk, and dairy. Because most vegetarian diets exclude these products, tryptophan supply is reduced in these diets, and tryptophan is also a limiting amino acid in our menus. It is used in the synthesis of neurotransmitters and, as such, its significant deficiency causes a specific depressive reaction [45,46]. There are ways of increasing the biological quality of the consumed plant protein through the organization of menus in a way that would supplement the content of protein products with the missing amino acids. One of the ways is to combine legume products with grain seeds in one meal, e.g., beans with rice. In comparison to animal proteins, plant proteins have lower contents of leucine, lysine, methionine, and tryptophan [42].

Vegetarian diets are associated with the lower level of cholesterol in the plasma and lower blood pressure. However, this is strongly associated with the lifestyle of vegetarians—these people are usually non-smokers, they do not drink alcohol, and are physically active.

The regular diet of Poles was characterized by the highest percentage of fat, unsaturated fatty acids, and cholesterol in comparison to other menus. This correlates with the increased risk of developing ischemic heart disease, atherosclerosis, and cancers—prostate, breast, or colon. Circulatory system diseases have been the main cause of death in the Polish population in recent decades [43]. The lowest content of PUFA was observed in FFD, RD, and PV. PUFA have a positive influence on the functioning of the circulatory system, e.g., they contribute to the decrease in the level of cholesterol, they lower arterial blood pressure, prevent the development of clots, and increase the strength of heart contraction [47].

High content of red meat, processed meat products, and eggs in these menus was the reason for the excessive consumption of cholesterol and saturated fatty acids, which are a significant factor contributing to deaths resulting from CVD. High level of cholesterol is also associated with the development of colorectal cancer [48]. Excessive levels of cholesterol also contribute to the development and progression of neurodegenerative diseases, such as Alzheimer's disease and Parkinson's disease [49]. Cholesterol is necessary for the proper development of the fetus in the first stages of pregnancy. After being born, about 40-50% of the child's cholesterol intake comes from mother's milk. Because of that, the VEGAN diet should not be recommended to pregnant or breastfeeding women [50].

Assimilable carbohydrates that are digested in the gastrointestinal tract are responsible for the supply of energy to muscles, the brain, heart, intestines, and erythrocytes. Dietary fiber (cellulose, hemicellulose, pectin, lignin) is not digested by enzymes in the gastrointestinal tract. The main functions of dietary fiber in the body are: the reduction of the level of cholesterol, glucose and insulin, the stimulation of fermentation processes in the intestine, the decrease in the time of intestinal passage, and the increase in the volume of stool [37]. According to WHO/FAO, the daily consumption of 25 g of fiber enables the correct functioning of the body. Basing on our original study, the lowest and insufficient content of fiber was observed in the RD. The insufficient supply of fiber in the diet may be associated with the development of disorders in the functioning of intestines, as well as with the increase in the risk of coronary artery disease and type 2 diabetes [51]. The main products of fermentation bacteria in the intestines are SCFA, especially acetate, propionate, and butyrate. They have many properties that are beneficial to health, they are responsible for feeling full, and they stimulate the immune system. In the case when there is a deficiency of dietary fiber, microbes shift to less favorable energy sources [52]. Moreover, prolonged consumption of high-fat and high-saccharose diet may lead to the death of the positive species of gut microflora [53]. Despite the numerous health benefits related to the high consumption of fiber, its excessive intake can have negative consequences. Products rich in food fiber, i.e., legumes, nuts, tofu, and some cereals are characterized by high content of phytic acid. Phytates may bind with some minerals, e.g., iron, zinc, and calcium, forming insoluble complexes, reducing their assimilation in the digestive tract [54]. This is why when using diets based on plant products (mainly legumes and cereals as in VEGAN and PV) it is important to control the levels of minerals in the body.

The regular diet was characterized by the highest content of saccharose out of all of the analyzed diets. This was caused by the presence of significant amounts of candy and sugar in the diet, the latter being added, e.g., to coffee, tea, and processed foods. Saccharose, which consists of glucose and fructose, also occurs naturally in honey, fruit, and vegetables, but in significantly lower quantities than in ready-made products prepared by the food industry. Excessive consumption of sugar is associated with many negative health aspects, such as circulatory system diseases, obesity, type 2 diabetes, caries, cirrhosis, and dementia [55]. Saccharose in the rest of the analyzed diets mainly originates from fruit, vegetables, and honey which, apart from being a source of sugar, include numerous valuable vitamins, minerals, as well as fiber.

The highest average content of lactose was observed in LOV, PV, and FFD. Very small amounts of milk sugar were observed in MFD and VEGAN, which originated from bread added in the dietetics program, though there should be no trace of this sugar in the menus at all. MFD and VEGAN could be appropriate for people with confirmed lactose intolerance and the incorrect assimilation of lactose in the digestive tract. Studies show that the risk of symptoms after the consumption of lactose depends on the dose of lactose, lactose expression, intestinal flora, and the sensitivity of the digestive tract [56]. When treating lactose intolerance, it is recommended to reduce lactose consumption, not eliminate it from the diet completely because in blind studies, most patients that reported the intolerance tolerated at least 12 g of lactose (which is equivalent to 250 mL of milk), and up to 18 g with the consumed foods [57]. A study conducted by Staudacher et al. regarding a diet poor in fermentable oligosaccharides, disaccharides, monosaccharides and polyols (FODMAP) indicated the improvement of symptoms in 86% of patients with the irritable bowel syndrome (IBS), in comparison to 49% in the case of a standard dietary intervention [58]. FODMAP is a diet that includes low contents of fermenting oligo-, di- and monosaccharides, as well as polyols, so fructose, lactose, fructans, galactans, and artificial sweeteners like sorbitol, mannitol, maltitol, and xylitol. All of these constituents are poorly absorbed in the small intestine, they are osmotically active (they can have laxative effects, they influence intestinal motility), and they are quickly fermented by intestinal bacteria. FODMAP aims at reducing or eliminating the presence of such symptoms as flatulence, stomachache, nausea, diarrhea, and constipation [59].

The regular diet of Poles was characterized by significantly higher consumption of sodium. Only VEGAN menus were characterized by lower levels of this constituent. High-sodium diet significantly increases the risk of developing hypertension, insulin resistance, dyslipidemia, and hipoadiponectemia [60]. The widespread supply of sodium is considered as one of the main causes of death resulting from circulatory system diseases. Sportsmen require higher intake of sodium, especially those that exercise in high temperatures. Higher sodium intake is also important for patients with insufficiency of the adrenal cortex and thyroid. During intense physical activity, contestants lose this element along with sweat [61]. Because of that, VEGAN may not be the right choice for people that do intense exercise.

Out of all of the analyzed diets, only RD was characterized by insufficient levels of potassium, when compared to the valid norm. Incorrect supply of this element may be associated with the increased risk of stroke and other circulatory system diseases [62]. Studies conducted by Zhang et al. indicate that excessive consumption of sodium positively correlated with increased systolic blood pressure and hypertension, and that the consumption of potassium negatively correlate with both of these disorders. Furthermore, the ratio of sodium and potassium was also important in the prevention of these problems [63].

The analysis of diets in terms of calcium content revealed that MFD, RD, and VEGAN contain insufficient amounts of this element. Well-absorbed sources of calcium are milk and milk products. Other sources include small fish (consumed with bones), beans, kale, parsley leaves, nuts, almonds, sesame seeds, and poppy seeds. It has to be highlighted that calcium originating from plant sources is less efficiently absorbed than calcium from milk and its products, which is associated with, e.g., the presence of lactose, which amplifies the absorption of this element [64]. However, it is emphasized that the best absorbed sources of calcium are vegetable with low in oxalate [65]. MFD completely eliminated milk products, this is why the average content of this macroelement was so low despite the use of other products that are its source. Due to the fact that the consumption of minerals with water was not taken into account, deficiencies in all types of diets, especially in terms of calcium, may be much smaller. The right supplementation of calcium and Vitamin D is of key importance for the prevention of the progressing loss of bone mass. In the case of postmenopausal women, it is recommended to consume a daily total of 1200 mg of calcium originating from food and supplements, and to supplement the diet with 800–2000 IU of Vitamin D. The supplementation is insufficient to prevent bone breaking in persons with osteoporosis. However, this is an important addition to a pharmacological intervention [66].

Phosphorus deficiencies were not observed in the studied diets. However, the proportion of Ca:P should be 1:1 to maintain the proper state of the skeleton. Mineral metabolism dysfunctions are the frequent complications of chronic kidney disease (CKD). A damaged kidney is not able to fully dispose of a phosphorus charge, leading to, e.g., secondary hyperparathyroidism. Studies conducted by Moe et al. indicated that protein products rich in phosphorus, i.e., cereals and legumes, are a better source of protein for people with CKD. The results of this study show that the use of the vegetarian diet in patients with CKD leads to the reduction of the level of phosphorus in the serum, when compared to a diet that includes meat [67].

The analyzed RD did not fulfill the RDA norm for magnesium with reference to men. Studies conducted by Adebamowo et al. indicated that a magnesium-, potassium-, and calcium-rich diet may contribute to the decrease in the risk of stroke in men [68]. Generally speaking, no pathological states associated with low magnesium consumption have been observed, but a small to moderate deficiency of this element resulting from chronic stress may significantly contribute to the presence of such illnesses as atherosclerosis, hypertension, osteoporosis, diabetes, and cancer [69]. Coffee, which is often consumed in large amounts by the Polish population, is a factor that is commonly believed to decrease the assimilation of magnesium, which favors many pathologies [70,71].

Iron present in food products has many forms and is usually classified as heme and nonheme iron. All of the analyzed diets fulfilled RDA for iron, but they differed in terms of its origin. In the case of meat-eliminating diets, it is mainly nonheme iron, which can occur in products in the form of various complexes, which may improve or weaken its absorption. An example of substances that significantly reduce the absorption of iron in the digestive tract are phytates and tannins of plant origin [72]. This is why the content of nonheme iron in the diet should be several times higher than heme iron.

RDA norms for such minerals as zinc, copper, and manganese have been fulfilled by all of the analyzed diets. However, taking into account factors that interrupt absorption and assimilation in the digestive tract, the supply of these constituents may turn out to be too low. A frequent factor facilitating this process is animal protein, which is not present in VEGAN. It has been demonstrated that the supplementation with zinc has a protective effect on the epithelial barrier of intestines and helps in various pathologies, including chronic alcohol consumption, oxidative stress, diarrhea, chronic fatigue syndrome, colitis, other gastrointestinal problems, and even some neurological disorders. However, zinc deficiency may result from the wide use of proton-pump inhibitor medicines, diets including large amounts of products rich in phytates and the decreasing consumption of meat and fish [50,73].

The main sources of copper are food products (75%) and drinking water (25%). Genetic illnesses connected to the disturbed metabolism of this element include Menkes disease, associated with bad absorption, and Wilson's disease, in which the excretion of iron is disturbed. Infants are more vulnerable to the deficiency than adults. This is true especially for premature infants because the fetus absorbs copper in the last months of pregnancy. Children that do not go through breastfeeding require supplementation in the first year of their lives [74]. However, high copper consumption with trans-fat and saturated fatty acids has been associated with the accelerated decrease in cognitive functions of the elderly [75].

Manganese is a necessary element that is required for the proper functioning of the immune system, the regulation of the level of sugar in the blood, cellular energy, reproduction, digestion, bone growth, blood coagulation, hemostasis, and protection against reactive oxygen species. Manganese deficiencies are rarely observed because it is available in many food products. The absorption of manganese is strictly regulated in intestines and no toxicity resulting from its excessive intake with the diet was observed. The toxicity of manganese in the world results from environmental pollution, including the pollution of air and drinking water [76].

All of the analyzed diets were characterized by insufficient iodine content when compared to the norm. However, the addition of salt was not included in the meals. Thanks to fortification (in Poland in the form of potassium iodide), salt is the main source of iodine in the diet of Poles. The main sources of this constituent include sea fish, which were not present in FFD, LOV, VEGAN, and PV. This is why the average content of iodine in these diets is the lowest. The main results of the deficiency are goiter and hypothyroidism. In pregnant women, insufficient consumption of iodine may be associated with impaired psychomotor development of children, the risk of miscarriage or endemic cretinism. The impaired mental and somatic development may result from the deficiency of iodine in children and teenagers [77].

All of the analyzed diets fulfilled the norm for Vitamin A, sometimes significantly exceeding the recommended values. The lowest content of retinol was observed in VEGAN because the diet completely eliminates products of animal origin, which are its source. RD has the lowest average content of beta-carotene; a carotenoid present in plant products. The reason for the limited assimilation of Vitamin A may be the excessive intake of fiber or alcohol, excess amounts of iron, nitrates, nitrites, and free radicals in the body, as well as insufficient levels of zinc [41]. The symptoms of Vitamin A deficiency include weaker sight, skin dryness, the weakening of mucous membranes, higher vulnerability to infections.

The diets were properly balanced in terms of Vitamin E, only RD did not fulfill the RDA norm for men. Vitamin E also influences the efficiency of muscles and the production of sperm. Therefore, its appropriate supply is very important in men [78].

The effects of vitamins from the B family were observed in many aspects, including brain function, energy production, DNA and RNA synthesis and repair, as well as in the synthesis of numerous neurochemical substances and signaling particles. Insufficient amounts of B-family vitamins are associated with inflammatory processes and oxidative stress as indicated by the increased concentration of homocysteine in blood plasma [79]. The average content of B-family vitamins in the analyzed diets fulfilled the norm of the recommended consumption for the Polish population. The source of these vitamins includes both products of plant and animal origin. The only exception is Vitamin B_{12}, which can be found only in animal foods. The VEGAN diet achieved the norm for this Vitamin Because the products used in the menus, i.e., soy milk or tofu, were enriched. The group of B vitamins includes folates (Vitamin B_9). The regular diet of Poles turned out to be insufficient in terms of this nutrient. The methylenetetrahydrofolate reductase (MTHFR) 3 gene codes the methylene tetrahydrofolate reductase enzyme, which participates in the metabolism of folates, homocysteine, and methionine. MTHFR transforms folic acid from food into an active form, which can be used by the organism. This way, MTHFR influences the transformation of toxic homocysteine into methionine with the participation of folic acid. The presence of the C677T mutation of the MTHFR gene leads to the deficiency of folic acid and the accumulation of homocysteine. It is estimated that about 15% of the Polish population has the mutation of the MTHFR gene. More efficiently assimilated vitamins are those that originate from the diet, not supplements. The most valuable sources of B-family vitamins are meat, fish, seafood, nuts, liver, green leafy vegetables, yeast, and eggs. People with the MTHFR gene mutation should be supplemented with methylated folic acid because this is the only form that will be assimilated by the body. The C677T polymorphism of methylene tetrahydrofolate reductase is associated with various illnesses, i.e., circulatory system diseases, neoplasms, neurological diseases, diabetes, or psoriasis [80].

Vitamin C (ascorbic acid) has strong antioxidant properties, is well assimilated from the digestive tract, and its excess is removed with urine. The main sources of ascorbic acid in a diet are fruit and vegetables, but there are huge losses of this constituent during heat treatment and storage, even >75% when compared to fresh, raw product [41]. The regular diet (RD) was characterized by the lowest content of Vitamin C and, at the same time, it did not fulfill the norms of consumption. The remaining diets (MFD, FFD, BD, LOV, VEGAN, and PV) covered the norm for this vitamin. The symptoms of scurvy or Vitamin C deficiency include swelling of the lower limbs, bleeding of gums, tiredness, and hemorrhages, as well as psychological issues, including depression, hysteria, and social introversion [81].

Vitamin D deficiency is the main problem of public health in the entire world in all age groups, even in countries where it is generally assumed that UV radiation is sufficient to prevent this deficiency or in industrialized countries where fortification has been conducted for years.

The causes of Vitamin D deficiency:

- the use of sunscreen,
- elderly age,
- obesity,
- malabsorption,
- kidney and liver diseases,
- use of anticonvulsants.

Due to the fact that it is very difficult to supplement proper amounts of Vitamin D with food, none of the analyzed diets fulfilled the norm for this constituent. Regardless of whether the diet eliminated products of animal origin or not, supplementation is necessary, particularly in the autumn-winter period in the temperate climate that is present in Poland. It has been demonstrated that Vitamin D stimulates the absorption of calcium in the intestines [82]. Vitamin D deficiency is usually manifested through the deformation of bones (rickets) or hypocalcaemia in infancy and childhood, as well as through pain and musculoskeletal weakness in adults. Many other health problems, including circulatory system diseases, type 2 diabetes, several neoplasms, and autoimmune diseases, can be associated with Vitamin D deficiency [83].

The analysis of menus prepared by qualified dieticians will make it possible to avoid some deficiencies. However, it is important to take into account the fact that most people who follow this type of diet do it rather poorly, which—in most cases—leads to even higher deficiencies, including calorie deficiencies, especially in the VEGE diet. An important assumption of this study was the same caloric value for every type of the analyzed diets.

Moreover, dietary fiber plays a key role in many metabolic processes not only directly related to the function of the intestine. Vegetable fiber is used by the intestinal microbes (stimulating growth of intestinal microbes) to synthesize SCFA, which support healthy colonic epithelial cells [84]. Fiber consumption directly affects stool bulkiness, fecal pH, and intestinal transit time. The end products (acetate, propionate, and butyrate) produced by microorganisms affect enhancing various blood parameters (glucose, insulin) and the manner of bowel movements. The study showed that the numbers of bifidobacteria, lactobacilli, and methanogens were significantly decreased in the colon of patients with mixed refractory constipation [84,85]. Taking into account the above considerations, conclusions were drawn. Periodic implementation of vegetarian diets may help in the treatment of several diseases and symptoms associated with the metabolic syndrome, such as hypertension, hyperlipidemia, obesity, type 2 diabetes, and cardiovascular diseases. The implementation of diets that eliminate products of animal origin can be risky for pregnant or breastfeeding women, children, and the elderly.

6. Conclusions

The regular diet of Poles featured the highest total content of fats and the highest content of saturated fatty acids and cholesterol. The regular diet of Poles, at 2000 kcal caloricity, turned out to be more hazardous to health in terms of deficiencies than properly balanced diets with the same caloricity that eliminated products of animal origin. Considering assimilation capabilities, metabolism and the source of vitamins and minerals in individual menus, it was impossible to clearly determine a better way of fulfilling the requirements for these constituents in vegetarian diets. Diets that eliminated products of animal origin often require additional supplementation and the constant monitoring of mineral and vitamin levels in blood plasma.

It is important to pay attention to the nutritional education of Poles, both with reference to those that implement diets featuring all product groups, as well as those that follow vegetarian diets.

Supplementary Materials: The following are available online at http://www.mdpi.com/2072-6643/12/10/2986/s1, Table S1: Differences in mineral content, Table S2: The presence of statistically significant differences in the content of vitamins between the diets.

Author Contributions: Conceptualization, K.K., M.S.; methodology, K.K., M.S. software K.K., M.S.; validation K.K., J.B., M.S. formal analysis, K.K., M.S.; investigation, K.K., J.B., M.S. resources, K.K., J.B., M.S. data curation K.K., J.B., K.P.-N., M.S. writing—original draft preparation M.S.; writing review & editing—J.B., K.P.-N., M.S.; visualization K.P.-N., M.S.; supervision M.S.; project administration, M.S.; funding acquisition, J.B. M.S. All authors have read and agreed to the published version of the manuscript.

Funding: This research received no external funding.

Conflicts of Interest: The authors declare no conflict of interest.

Abbreviations

BD	The basic diet
CKD	chronic kidney disease
CVD	cardiovascular disease
DHA	Docosahexaenoic Acid
EPA	Eicosapentaenoic acid
FFD	Diet that eliminates fish, seafood and shellfish
FOODMAP	fermentable oligosaccharides, disaccharides, monosaccharides and polyols
LDL	Low-density lipoprotein
LOV	The lacto-ovo-vegetarian diet
MFD	The milk-free diet
MTHFR	methylenetetrahydrofolate reductase

MUFA	monounsaturated fatty acids
PUFA	polyunsaturated fatty acids
RD	regular diet of Poles
SFA	saturated fatty acids
VEGAN	The vegan diet
PV	The modification of a vegetarian diet with fish

References

1. Milton, K. The critical role played by animal source foods in human (Homo) evolution. *J. Nutr.* **2003**, *133*, 3886S–3892S. [CrossRef] [PubMed]
2. Singh, V.; Neelam, S. Meat Species Specifications to Ensure the Quality of Meat—A Review. *Int. J. Meat Sci.* **2011**, *1*, 15–26. [CrossRef]
3. Ruby, M.B. Vegetarianism. A blossoming field of study. *Appetite* **2012**, *58*, 141–150. [CrossRef] [PubMed]
4. World Cancer Research Fund International, Cancer Prevention Recommendations. Available online: https://www.wcrf.org/dietandcancer/recommendations/limit-red-processed-meat (accessed on 21 November 2018).
5. McEvoy, C.T.; Temple, N.; Woodside, J.V. Vegetarian diets, low-meat diets and health: A review. *Public Health Nutr.* **2012**, *15*, 2287–2294. [CrossRef] [PubMed]
6. Ferdowsian, H.R.; Barnard, N.D. Effects of Plant-Based Diets on Plasma Lipids. *Am. J. Cardiol.* **2009**, *104*, 947–956. [CrossRef]
7. Fontana, L.; Klein, S. Aging, Adiposity, and Calorie Restriction. *JAMA* **2007**, *297*, 986. [CrossRef]
8. Pobłocki, J.; Jasińska, A.; Syrenicz, A.; Andrysiak-Mamos, E.; Szczuko, M. The Neuroendocrine Neoplasms of the Digestive Tract: Diagnosis, Treatment and Nutrition. *Nutrients* **2020**, *12*, 1437. [CrossRef]
9. Surh, Y.-J. Cancer chemoprevention with dietary phytochemicals. *Nat. Rev. Cancer* **2003**, *3*, 768–780. [CrossRef]
10. Barrea, L.; Altieri, B.; Muscogiuri, G.; Laudisio, D.; Annunziata, G.; Colao, A.; Faggiano, A.; Savastano, S. Impact of Nutritional Status on Gastroenteropancreatic Neuroendocrine Tumors (GEP-NET) Aggressiveness. *Nutrients* **2018**, *10*, 1854. [CrossRef]
11. Wang, X.; Ouyang, Y.; Liu, J.; Zhu, M.; Zhao, G.; Bao, W.; Hu, F.B. Fruit and vegetable consumption and mortality from all causes, cardiovascular disease, and cancer: Systematic review and dose-response meta-analysis of prospective cohort studies. *BMJ* **2014**, *349*, g4490. [CrossRef]
12. Omyła-Rudzka, M. *Centre for Public Opinion Research*; Research Report CBOS: Warszawa, Poland, 2014; p. 113, ISSN 2353-5822.
13. Fraser, G.E. Vegetarian diets: What do we know of their effects on common chronic diseases? *Am. J. Clin. Nutr.* **2009**, *89*, 1607S–1612S. [CrossRef] [PubMed]
14. Tonstad, S.; Butler, T.; Yan, R.; Fraser, G.E. Type of vegetarian diet, body weight, and prevalence of type 2 diabetes. *Diabetes Care* **2009**, *32*, 791–796. [CrossRef] [PubMed]
15. Mozaffarian, D.; Micha, R.; Wallace, S. Effects on Coronary Heart Disease of Increasing Polyunsaturated Fat in Place of Saturated Fat: A Systematic Review and Meta-Analysis of Randomized Controlled Trials. *PLoS Med.* **2010**, *7*, e1000252. [CrossRef]
16. Wu, A.H.; Koh, W.-P.; Wang, R.; Lee, H.-P.; Yu, M.C. Soy intake and breast cancer risk in Singapore Chinese Health Study. *Br. J. Cancer* **2008**, *99*, 196–200. [CrossRef]
17. Iannotti, L.L.; Lutter, C.K.; Bunn, D.A.; Stewart, C.P. Eggs: The uncracked potential for improving maternal and young child nutrition among the world's poor. *Nutr. Rev.* **2014**, *72*, 355–368. [CrossRef] [PubMed]
18. Kijowski, J.; Leśnierowski, G.; Cegielska-Radziejewska, R. Eggs are a valuable source of bioactive ingredients, FOOD. *Sci. Technol. Qual.* **2013**, *5*, 29–41.
19. Plotnikoff, G.A. Nutritional assessment in vegetarians and vegans: Questions clinicians should ask. *Minn. Med.* **2012**, *95*, 36–38.
20. Gilsing, A.M.; Crowe, F.L.; Lloyd-Wright, Z.; Sanders, T.A.; Appleby, P.N.; Allen, N.E.; Key, T.J. Serum concentrations of Vitamin B12 and folate in British male omnivores, vegetarians and vegans: Results from a cross-sectional analysis of the EPIC-Oxford cohort study. *Eur. J. Clin. Nutr.* **2010**, *64*, 933–939. [CrossRef]
21. Stabler, S.P. Vitamin B12 Deficiency. *N. Engl. J. Med.* **2013**, *368*, 149–160. [CrossRef]

22. Le, L.T.; Sabaté, J. Beyond Meatless, the Health Effects of Vegan Diets: Findings from the Adventist Cohorts. *Nutrients* **2014**, *6*, 2131–2147. [CrossRef]

23. Jenkins, D.J.A.; Kendall, C.W.C.; Augustin, L.S.A.; Mitchell, S.; Sahye-Pudaruth, S.; Mejia, S.B.; Chiavaroli, L.; Mirrahimi, A.; Ireland, C.; Bashyam, B.; et al. Effect of Legumes as Part of a Low Glycemic Index Diet on Glycemic Control and Cardiovascular Risk Factors in Type 2 Diabetes Mellitus. *Arch. Intern. Med.* **2012**, *172*, 1653–1660. [CrossRef] [PubMed]

24. Mattar, R.; Mazo, D.F.D.C.; Carrilho, F.J. Lactose intolerance: Diagnosis, genetic, and clinical factors. *Clin. Exp. Gastroenterol.* **2012**, *5*, 113–121. [CrossRef]

25. Lomer, M.; Parkes, G.C.; Sanderson, J.D. Review article: Lactose intolerance in clinical practice—Myths and realities. *Aliment. Pharmacol. Ther.* **2007**, *27*, 93–103. [CrossRef] [PubMed]

26. Ohlsson, J.A.; Johansson, M.; Hansson, H.; Abrahamson, A.; Byberg, L.; Smedman, A.; Lindmark-Månsson, H.; Lundh, Å. Lactose, glucose and galactose content in milk, fermented milk and lactose-free milk products. *Int. Dairy J.* **2017**, *73*, 151–154. [CrossRef]

27. Szczuko, M.; Ziętek, M.; Kulpa, D.; Seidler, T. Riboflavin—Properties, occurrence and its use in medicine. *Pteridines* **2019**, *30*, 33–47. [CrossRef]

28. Muehlhoff, E.; Bennett, A.; McMahon, D. *Milk and Dairy Products in Human Nutrition [Online]*; Food and Agriculture Organization of the United Nations (FAO): Rome, Italy 2013; ISBN: 9789251078631. Available online: http://www.fao.org/docrep/018/i3396e/i3396e.pdf (accessed on 9 December 2018).

29. Danby, F.W. Diet in the prevention of hidradenitis suppurativa (acne inversa). *J. Am. Acad. Dermatol.* **2015**, *73*, S52–S54. [CrossRef]

30. Lazzarin, N.; Vaquero, E.; Exacoustos, C.; Bertonotti, E.; Romanini, M.E.; Arduini, D. Low-dose aspirin and omega-3 fatty acids improve uterine artery blood flow velocity in women with recurrent miscarriage due to impaired uterine perfusion. *Fertil. Steril.* **2009**, *92*, 296–300. [CrossRef]

31. Swanson, D.; Block, R.; Mousa, S.A. Omega-3 Fatty Acids EPA and DHA: Health Benefits Throughout Life1. *Adv. Nutr.* **2012**, *3*, 1–7. [CrossRef]

32. Domingo, J.L. Nutrients and Chemical Pollutants in Fish and Shellfish. Balancing Health Benefits and Risks of Regular Fish Consumption. *Crit. Rev. Food Sci. Nutr.* **2014**, *56*, 979–988. [CrossRef]

33. Copat, C.; Arena, G.; Fiore, M.; Ledda, C.; Fallico, R.; Sciacca, S.; Ferrante, M. Heavy metals concentrations in fish and shellfish from eastern Mediterranean Sea: Consumption advisories. *Food Chem. Toxicol.* **2013**, *53*, 33–37. [CrossRef]

34. Dobrzyńska, M.; Przysławski, J. Prevention of cardiovascular disease and eating behavior in group of women and men aged 20 to 30 years. *J. Med. Sci.* **2014**, *83*, 116–121.

35. Gil, M.; Glodek, E.; Rudy, M. Assessment of the consumption of vitamins and minerals in daily food rations of students of the University of Rzeszów. *Ann. Natl. Inst. Hyg.* **2012**, *63*, 441–446.

36. Szczuko, M.; Gutowska, I.; Seidler, T. Nutrition and nourishment status of Polish students in comparison with students from other countries. *Roczniki Państwowego Zakładu Higieny* **2015**, *66*, 261–268. [PubMed]

37. Jarosz, M. *Normy żywienia dla populacji Polski*; Instytut Żywności i Żywienia: Warsaw, Poland, 2017.

38. Richter, C.K.; Skulas-Ray, A.C.; Champagne, C.M.; Kris-Etherton, P.M. Plant protein and animal proteins: Do they differentially affect cardiovascular disease risk? *Adv. Nutr.* **2015**, *6*, 712–728. [CrossRef] [PubMed]

39. Bernstein, A.M.; Sun, Q.; Hu, F.B.; Stampfer, M.J.; Manson, J.E.; Willett, W.C. Major Dietary Protein Sources and Risk of Coronary Heart Disease in Women. *Circulation* **2010**, *122*, 876–883. [CrossRef] [PubMed]

40. Sun, Q.; Pan, A.; Bernstein, A.M.; Schulze, M.B.; Manson, J.E.; Stampfer, M.J.; Willett, W.C.; Hu, F.B. Red Meat Consumption and Mortality: Results from 2 prospective cohort studies. *Arch. Intern. Med.* **2012**, *172*, 555–563. [CrossRef]

41. Włodarek, D.; Lange, E.; Kozłowska, L.; Głąbska, D. *Dietotherapy*; PZWL Medical: Warsaw, Poland, 2015; ISBN 978-83-200-4861-2.

42. Marsh, K.; Munn, E.A.; Baines, S.K. Protein and vegetarian diets. *Med J. Aust.* **2013**, *199*, S7–S10. [CrossRef]

43. Krajcovicová-Kudláčková, M.; Babinska, K.; Valachovicova, M. Health benefits and risks of plant proteins. *Bratisl Lek List.* **2005**, *106*, 231–234.

44. Thorpe, D.L.; Knutsen, S.F.; Beeson, W.L.; Rajaram, S.; Fraser, G.E. Effects of meat consumption and vegetarian diet on risk of wrist fracture over 25 years in a cohort of peri- and postmenopausal women. *Public Health Nutr.* **2008**, *11*, 564–572. [CrossRef]

45. Booij, L.; Van Der Does, W.; Haffmans, P.M.J.; Spinhoven, P.; McNally, R.J. Acute tryptophan depletion as a model of depressive relapse. *Br. J. Psychiatry* **2005**, *187*, 148–154. [CrossRef]

46. Hibbeln, J.R.; Northstone, K.; Evans, J.; Golding, J. Vegetarian diets and depressive symptoms among men. *J. Affect. Disord.* **2018**, *225*, 13–17. [CrossRef] [PubMed]

47. Cierniak-Piotrowska, M.; Marciniak, G.; Stańczak, J. *Morbidity and Mortality from Cardiovascular Diseases and the Demographic Situation of Poland*; Central Statistical Office: Delhi, India, 2015.

48. Yan, G.; Li, L.; Zhu, B.; Li, Y. Lipidome in colorectal cancer. *Oncotarget* **2016**, *7*, 33429–33439. [CrossRef] [PubMed]

49. Maulik, M.; Westaway, D.; Jhamandas, J.H.; Kar, S. Role of Cholesterol in APP Metabolism and Its Significance in Alzheimer's Disease Pathogenesis. *Mol. Neurobiol.* **2012**, *47*, 37–63. [CrossRef] [PubMed]

50. Baardman, M.E.; Kerstjens-Frederikse, W.S.; Berger, R.M.; Bakker, M.K.; Hofstra, R.M.; Plösch, T. The Role of Maternal-Fetal Cholesterol Transport in Early Fetal Life: Current Insights1. *Boil. Reprod.* **2013**, *88*, 24. [CrossRef]

51. Mudgil, D.; Barak, S. Composition, properties and health benefits of indigestible carbohydrate polymers as dietary fiber: A review. *Int. J. Boil. Macromol.* **2013**, *61*, 1–6. [CrossRef]

52. Wall, R.; Ross, R.P.; Shanahan, F.; O'Mahony, L.; O'Mahony, C.; Coakley, M.; Hart, O.; Lawlor, P.; Quigley, E.M.; Kiely, B.; et al. Metabolic activity of the enteric microbiota influences the fatty acid composition of murine and porcine liver and adipose tissues. *Am. J. Clin. Nutr.* **2009**, *89*, 1393–1401. [CrossRef]

53. Sonnenburg, E.D.; Smits, S.A.; Tikhonov, M.; Higginbottom, S.K.; Wingreen, N.S.; Sonnenburg, J.L. Diet-induced extinctions in the gut microbiota compound over generations. *Nature* **2016**, *529*, 212–215. [CrossRef]

54. Skrovanek, S.; DiGuilio, K.; Bailey, R.; Huntington, W.; Urbas, R.; Mayilvaganan, B.; Mercogliano, G.; Mullin, J.M. Zinc and gastrointestinal disease. *World J. Gastrointest. Pathophysiol.* **2014**, *5*, 496–513. [CrossRef]

55. Schmidt, L.A. New unsweetened truths about sugar. *JAMA Intern. Med.* **2014**, *174*, 525. [CrossRef]

56. Misselwitz, B.; Pohl, D.; Frühauf, H.; Fried, M.; Vavricka, S.R.; Fox, M. Lactose malabsorption and intolerance: Pathogenesis, diagnosis and treatment. *United Eur. Gastroenterol. J.* **2013**, *1*, 151–159. [CrossRef]

57. Shaukat, A.; Levitt, M.D.; Taylor, B.C.; Macdonald, R.; Shamliyan, T.; Kane, R.L.; Wilt, T.J. Systematic Review: Effective Management Strategies for Lactose Intolerance. *Ann. Intern. Med.* **2010**, *152*, 797–803. [CrossRef]

58. Staudacher, H.M.; Whelan, K.; Irving, P.M.; Lomer, M. Comparison of symptom response following advice for a diet low in fermentable carbohydrates (FODMAPs) versus standard dietary advice in patients with irritable bowel syndrome. *J. Hum. Nutr. Diet.* **2011**, *24*, 487–495. [CrossRef] [PubMed]

59. Gibson, P.R.; Shepherd, S.J. Evidence-based dietary management of functional gastrointestinal symptoms: The FODMAP approach. *J. Gastroenterol. Hepatol.* **2010**, *25*, 252–258. [CrossRef] [PubMed]

60. Baudrand, R.; Campino, C.; Carvajal, C.A.; Olivieri, O.; Guidi, G.; Faccini, G.; Vöhringer, P.A.; Cerda, J.; Owen, G.; Kalergis, A.; et al. High sodium intake is associated with increased glucocorticoid production, insulin resistance and metabolic syndrome. *Clin. Endocrinol.* **2013**, *80*, 677–684. [CrossRef] [PubMed]

61. Jeukendrup, A.; Gleeson, M. *Sport Nutrition*, 3rd ed.; Human Kinetics Publishers: Windsor, ON, Canada, 2018; ISBN 978-1492529033.

62. D'Elia, L.; Iannotta, C.; Sabino, P.; Ippolito, R. Potassium-rich diet and risk of stroke: Updated meta-analysis. *Nutr. Metab. Cardiovasc. Dis.* **2014**, *24*, 585–587. [CrossRef] [PubMed]

63. Zhang, Z.; Cogswell, M.E.; Gillespie, C.; Fang, J.; Loustalot, F.; Dai, S.; Carriquiry, A.L.; Kuklina, E.V.; Hong, Y.; Merritt, R.; et al. Association between Usual Sodium and Potassium Intake and Blood Pressure and Hypertension among U.S. Adults: NHANES 2005–2010. *PLoS ONE* **2013**, *8*, e75289. [CrossRef] [PubMed]

64. Kwak, H.-S.; Lee, W.-J.; Lee, M.-R. Revisiting lactose as an enhancer of calcium absorption. *Int. Dairy J.* **2012**, *22*, 147–151. [CrossRef]

65. Agnoli, C.; Baroni, L.; Bertini, I.; Ciappellano, S.; Fabbri, A.; Papa, M.; Pellegrini, N.; Sbarbati, R.; Scarino, M.; Siani, V.; et al. Position paper on vegetarian diets from the working group of the Italian Society of Human Nutrition. *Nutr. Metab. Cardiovasc. Dis.* **2017**, *27*, 1037–1052. [CrossRef]

66. Khan, A.; Fortier, M. Osteoporosis in Menopause. *J. Obstet. Gynaecol. Can.* **2014**, *36*, 839–840. [CrossRef]

67. Moe, S.M.; Zidehsarai, M.P.; Chambers, M.A.; Jackman, L.A.; Radcliffe, J.S.; Trevino, L.L.; Donahue, S.E.; Asplin, J.R. Vegetarian Compared with Meat Dietary Protein Source and Phosphorus Homeostasis in Chronic Kidney Disease. *Clin. J. Am. Soc. Nephrol.* **2010**, *6*, 257–264. [CrossRef]

68. Adebamowo, S.N.; Spiegelman, D.; Flint, A.J.; Willett, W.C.; Rexrode, K.M. Intakes of Magnesium, Potassium, and Calcium and the Risk of Stroke among Men. *Int. J. Stroke* **2015**, *10*, 1093–1100. [CrossRef] [PubMed]

69. Nielsen, F.H. Magnesium, inflammation, and obesity in chronic disease. *Nutr. Rev.* **2010**, *68*, 333–340. [CrossRef] [PubMed]

70. Wierzejska, R.; Jarosz, M. Coffee drinking and risk of type 2 diabetes mellitus. Optimistic scientific data. *Przeglad Epidemiologiczny* **2012**, *66*, 509–512. [PubMed]

71. Dalton, L.M.; Fhloinn, D.M.N.; Gaydadzhieva, G.T.; Mazurkiewicz, O.M.; Leeson, H.; Wright, C.P. Magnesium in pregnancy. *Nutr. Rev.* **2016**, *74*, 549–557. [CrossRef]

72. Fuqua, B.K.; Vulpe, C.D.; Anderson, G.J. Intestinal iron absorption. *J. Trace Elem. Med. Boil.* **2012**, *26*, 115–119. [CrossRef]

73. Zou, T.-T.; Mou, J.; Zhan, X. Zinc Supplementation in Acute Diarrhea. *Indian J. Pediatr.* **2014**, *82*, 415–420. [CrossRef]

74. Mistry, H.D.; Kurlak, L.O.; Young, S.D.; Briley, A.L.; Pipkin, F.B.; Baker, P.N.; Poston, L. Maternal selenium, copper and zinc concentrations in pregnancy associated with small-for-gestational-age infants. *Matern. Child Nutr.* **2012**, *10*, 327–334. [CrossRef]

75. Squitti, R.; Siotto, M.; Polimanti, R. Low-copper diet as a preventive strategy for Alzheimer's disease. *Neurobiol. Aging* **2014**, *35*, S40–S50. [CrossRef]

76. Aschner, M.; Erikson, K. Manganese. *Adv. Nutr.* **2017**, *8*, 520–521. [CrossRef]

77. Gietka-Czernel, M. Prophylaxis of iodine deficiency. *Adv. Med Sci.* **2015**, *28*, 839–845.

78. Jiang, Q. Natural forms of Vitamin E: Metabolism, antioxidant, and anti-inflammatory activities and their role in disease prevention and therapy. *Free Radic. Boil. Med.* **2014**, *72*, 76–90. [CrossRef] [PubMed]

79. Kennedy, D.O. B Vitamins and the Brain: Mechanisms, Dose and Efficacy—A Review. *Nutrients* **2016**, *8*, 68. [CrossRef] [PubMed]

80. Liew, S.-C.; Das Gupta, E. Methylenetetrahydrofolate reductase (MTHFR) C677T polymorphism: Epidemiology, metabolism and the associated diseases. *Eur. J. Med. Genet.* **2015**, *58*, 1–10. [CrossRef] [PubMed]

81. Schlueter, A.K.; Johnston, C.S. Vitamin C: Overview and Update. *J. Evid. Based Integr. Med.* **2011**, *16*, 49–57. [CrossRef]

82. Holick, M.F.; Binkley, N.; Bischoff-Ferrari, H.A.; Gordon, C.M.; Hanley, D.A.; Heaney, R.P.; Murad, M.; Weaver, C.M. Evaluation, Treatment, and Prevention of Vitamin D Deficiency: An Endocrine Society Clinical Practice Guideline. *J. Clin. Endocrinol. Metab.* **2011**, *96*, 1911–1930. [CrossRef]

83. Pearce, S.H.S.; Cheetham, T.D. Diagnosis and management of Vitamin D deficiency. *BMJ* **2010**, *340*, b5664. [CrossRef]

84. Makki, K.; Deehan, E.C.; Walter, J.; Bäckhed, F. The Impact of Dietary Fiber on Gut Microbiota in Host Health and Disease. *Cell Host Microbe* **2018**, *23*, 705–715. [CrossRef]

85. So, D.; Whelan, K.; Rossi, M.; Morrison, M.; Holtmann, G.J.; Kelly, J.T.; Shanahan, E.R.; Staudacher, H.M.; Campbell, K.L. Dietary fiber intervention on gut microbiota composition in healthy adults: A systematic review and meta-analysis. *Am. J. Clin. Nutr.* **2018**, *107*, 965–983. [CrossRef]

© 2020 by the authors. Licensee MDPI, Basel, Switzerland. This article is an open access article distributed under the terms and conditions of the Creative Commons Attribution (CC BY) license (http://creativecommons.org/licenses/by/4.0/).

Article

Efficacy and Safety of Kudzu Flower–Mandarin Peel on Hot Flashes and Bone Markers in Women during the Menopausal Transition: A Randomized Controlled Trial

Ji Eon Kim [1], Hyeyun Jeong [1], Soohee Hur [1], Junho Lee [2] and Oran Kwon [1,*]

[1] Department of Nutritional Science and Food Management, Ewha Womans University, Seoul 03760, Korea; wldjs0903@naver.com (J.E.K.); jhy1630@hanmail.net (H.J.); soohee1276@naver.com (S.H.)
[2] LG Household and Healthcare Research Park, Seoul 07795, Korea; leejh222@lgcare.com
* Correspondence: orank@ewha.ac.kr; Tel.: +82-2-3277-6860

Received: 21 September 2020; Accepted: 20 October 2020; Published: 22 October 2020

Abstract: This randomized controlled study aimed to assess the efficacy and safety of an extract mixture of kudzu flower and mandarin peel (KM) on hot flashes (HFs) and markers of bone turnover in women during the menopausal transition. Healthy women aged 45–60 years with the menopausal HFs were randomly assigned in a 1:1 ratio to either KM (1150 mg/day) or placebo arms for 12 weeks ($n = 84$). The intent-to-treat analysis found that compared with the placebo, the KM significantly attenuated HF scores ($p = 0.041$) and HF severities ($p < 0.001$), with a mean difference from baseline to week 12. The KM also improved bone turnover markers, showing a significant reduction in bone resorption CTx ($p = 0.027$) and a tendency of increasing bone formation OC relative to the placebo. No serious adverse events and hormonal changes were observed in both groups. These findings suggest that KM consumption may improve the quality of life in ways that are important to symptomatic menopausal women.

Keywords: *Pueraria thomsonii*; *Citrus unshiu*; menopause; hot flash; bone resorption

1. Introduction

Menopause is a natural process of women aging that increases vulnerability to physical [1] and emotional [2] stresses, leading to a reduced quality of life and increased burden for health care needs [3,4]. From an endocrinological perspective, menopause is characterized by increased follicle-stimulating hormone (FSH) levels and a fluctuating/eventual decline in estrogen [5]. At some time through the menopausal transition from peri- to post-menopausal state, >80% of women will experience the vasomotor symptoms (VMS) that are commonly called hot flashes or flushes (HFs) and night sweats [6]. VMS peak in late perimenopause or early menopause due to the changes in circulating estrogen levels [7]. Meanwhile, because circulating estrogen plays a vital role in preventing bone loss and promoting bone formation, menopausal estrogen deficiency may also contribute to diminished bone mineral density (BMD) [8]. In a retrospective study, Crandall et al. [9] noted that lumbar BMD was lower in perimenopausal women with frequent HFs than in non-flushing women, supporting the notion that HFs may serve as an independent determinant of BMD. In contrast, Tuomikoeki et al. [10] suggested that HFs do not appear to determine lumbar and hip BMD in a prospective longitudinal study of 143 healthy women with or without HFs.

Hormone therapy has long been recognized as a standard treatment to reduce unpleasant symptoms of menopause. However, because of the risks of breast cancer, endometrial cancer, and venous thromboembolism [11], many women now want to avoid hormone therapy. Instead, various plant food products with estrogenic activity (phytoestrogens) are widely thought to confer health benefits to menopausal women. Globally, the phytoestrogen market was estimated to be valued at USD 1.2 billion in 2019, with a compound annual growth rate of 4.7% [12]. However, the prominent dietary and herbal sources of phytoestrogens are only partially known to be safe, effective, and well-tolerated by symptomatic menopausal women in controlled clinical trials. Therefore, there is still an unmet need for safe and effective non-hormone treatment options for menopausal women.

Kudzu (*Pueraria thomsonii* Benth.) belongs to the Fabaceae/Leguminosae family. The flower of kudzu, also called the Pueraria flower, contains tectorigenin, tectoridin, and tectorigenin 7-O-xylosylglucoside that have lower binding affinities to estrogen receptors (ERs) than those of soy isoflavones [13]. Mandarin (*Citrus unshiu* Markovich) is a popular citrus fruit belonging to the Rosaceae family. Dried mandarin peel is rich in flavonoids, such as hesperidin, naringin, and narirutin [14]. Current research has identified the modulatory effects of mandarin peel on bone metabolism [15] and inhibitory effects of hesperidin against bone loss in ovariectomized (OVX) animals [16]. Two recent animal studies focused on the HFs and BMD [17,18], respectively, in OVX mice treated with an extract mixture of kudzu flower and mandarin peel (KM) at the ratio of 6.5:5. The first study implied that a 7-week treatment with KM exerted protective effects against HFs by enhancing serotonin, norepinephrine, dopamine, and ERβ in the hypothalamus, and decreasing circulating FSH and luteinizing hormone (LH) levels without changing the circulating estrogen level [17]. In the second study, KM demonstrated anti-osteoporotic effects after a 7-week treatment by maintaining bone homeostasis via re-addressing the balance between bone resorption and bone formation [18]. However, to the best of our knowledge, KM's protective attributes have not yet been confirmed in human studies. It triggered us to study the potentials of KM in human subjects.

Based on these findings, we hypothesized that KM might potentially modify HFs and BMDs in menopausal women. To test this hypothesis, we conducted a double-blind placebo-controlled human trial to determine the safety and effectiveness of KM on modulating HFs and bone turnover markers in peri- and post-menopausal women.

2. Materials and Methods

2.1. Test Materials

The mandarin peel and kudzu flower are listed as Old Dietary Ingredients marketed in the US before 15 October 1994 [19]. Therefore, they might be used legally in a dietary supplement without submitting the Food and Drug Administration notification. The test and placebo products were provided by the LG Household & Health Care (Seoul, Korea). In brief, the 70% ethanolic extract of kudzu flower and aqueous extract of mandarin peel were spray-dried separately and mixed at the ratio of 6.5:5. The KM, lactose, crystalline cellulose, sodium carboxymethyl starch, silicon dioxide, magnesium stearate, and cacao color were packed in a capsule to provide a dose of 383.4 mg KM. For the placebo capsule, KM was replaced with an equal weight of lactose. The test product was standardized to contain 70.3 mg/g of three tectorigenin derivatives from the kudzu flower and 51.7 mg/g hesperidin from mandarin peel using a high-performance liquid chromatograph equipped with a variable wavelength detector (Agilent Technologies, Santa Clara, CA, USA) and Poroshell EC-C18 column (2.1 mm × 1000 mm, 2.7 μm).

2.2. Subjects

The required sample size was estimated based on data from Chang et al. [20] and Kim et al. [20,21] and conducted assuming a power of 0.80 with a two-sided α-level of 0.05 and 20% dropout rate. Subjects were recruited through posters placed in various locations and online advertisements. For inclusion, subjects were required to be peri- or post-menopausal women aged between 45 and 60 years with the menopausal HFs as indicated by a Kupperman Index (KI) ≥15 and average daily HF score ≥10 for 1 week before the screening visit. Exclusion criteria were as follows: (1) surgical or chemotherapy-induced menopause or unexplained vaginal bleeding; (2) receiving hormone therapy within the preceding 6 months; (3) taking a supplement or traditional medicine affecting menopausal status, bone health, blood glucose, blood lipids, blood pressure, and blood circulation within the preceding 1 month; (4) taking medication within the past 1 month or during the study period; (5) a history of breast cancer, endometrial hyperplasia, uterine endometrial cancer, sex steroid-dependent organ tumors, or having abnormal findings on breast X-ray or pelvic ultrasonography; (6) having liver or kidney diseases, uncontrolled hypertension (systolic blood pressure >160 mmHg or diastolic blood pressure >100 mmHg), thyroid disease, diabetes, hyperlipidemia, or mental illness; (7) a history of severe migraine headache, thromboembolic disorders, cerebrovascular disorders, or serious cardiovascular condition within the preceding 1 year; (8) drug addiction or alcoholism; (9) performing heavy exercise ($\geq$10 h/week) within the past 3 months; (10) known allergy or sensitivity to ingredients in the study products; (11) enrollment in any other clinical trial within the preceding 1 month; (12) investigator's determination of unsuitability for the trial. All participants provided written informed consent, and the Institutional Review Board of Ewha Womans University approved the protocol (IRB No. 164-14). The study was registered on the International Clinical Trials Registry Platform of the WHO (ICTRP) under the identification number KCT0003335.

2.3. Study Design

The study followed a randomized controlled parallel design with two arms of a fixed-dose of KM (1150 mg) or matching placebo in a ratio of 1:1. A total of 84 eligible subjects were randomized to the KM (n = 42) or placebo (n = 42) groups using a computer-generated random block number table. The group allocation was blinded for both the investigators and participants. The subjects were advised to take three test material caps with enough water once a day for 12 weeks. We determined the daily dose based on previous animal studies [18,22] and the duration by some previous studies [23]. During the trial, subjects were instructed to maintain their usual diet and lifestyle but refrain from eating or drinking kudzu, kudzu flower, mandarin peel, citrus fruits, calcium, and isoflavone-rich foods and beverages. Dietary intakes and responses to the International Physical Activity Questionnaire (IPAQ) [24] were recorded for 3 days (two weekdays and one weekend day) at baseline and weeks 6 and 12, using a smartphone application. Pittsburgh Sleep Quality Index (PSQI) [25] and Recommended Food Scores (RFS) [26] were measured at baseline and week 12 to assess the qualities of sleep and food intake, respectively. Blood samples were taken at baseline and week 12, after an overnight fast.

2.4. Measurement of HFs

Participants were required to complete a daily diary over 12 weeks from 7 days before treatment at baseline using a Web-based prospective electronic digital HF diary. The data were collected by the Google form survey method using a designed questionnaire in which participants recorded both the frequency and severity of HFs. For the frequency of HF, the daily number of episodes is recorded as one point. The severity of HF was defined as mild = 1, moderate = 2, severe = 3, and very severe = 4. The HF scores were calculated by multiplying the frequency experienced at that severity level. Then, the resulting points were summed to obtain a total daily score. Each daily score was averaged per week [27,28]. Daytime and night-time HFs were considered separately.

The frequency and severity of menopausal symptoms were also measured via the KI and Menopause-Specific Quality of Life Questionnaire (MenQOL) at baseline and end of the study. KI is a numerical index that scores 11 menopausal-related symptoms (HFs, paresthesia, insomnia, nervousness, melancholia, vertigo, weakness, arthralgia, myalgia, headache, and palpitations). Each symptom was rated from 0 to 3 according to severity and symptoms, weighted, and the total sum was calculated. MenQOL is a menopause-specific tool to measure health-related quality of life. It consists of a total of 29 items in a Likert-scale format, assessing the impacts on the four domains, including vasomotor (1–3 items), psychosocial (4–10 items), physical (11–26 items), and sexual (27–29 items) [29]. Means were computed by dividing the sum of the domain's items by the number of items within that domain.

2.5. Measurement of Biochemical Markers in the Blood

At baseline and week 12, venous blood was collected in the ethylenediaminetetraacetic acid tube (BD Biosciences, San Jose, CA, USA) and serum-separated tube (BD Biosciences). The plasma was separated by centrifugation at $1500 \times g$, 4 °C for 10 min, and serum was centrifuged at $1910 \times g$, 4 °C, for 15 min. Serum osteocalcin (OC) was measured by electrochemiluminescence (Elecsys N-MID Osteocalcin ELISA Kit, Roche Diagnostics GmbH, Mannheim, Germany). Plasma C-telopeptide fragment (CTx), plasma N-telopeptide fragment (NTx), and serum bone-specific alkaline phosphatase (BALP) were assessed by the Elecsys β-CrossLaps/serum assay (Roche Diagnostics GmbH), enzyme-linked immunosorbent assay kit (Cusabio Biotech, Wuhan, China), and the Access Ostase assay (Beckman Coulter, Fullerton, CA, USA), respectively. Estradiol, FSH, and LH levels were determined by the Cobas e801 analyzer (Roche Diagnostics GmbH).

2.6. Safety Measurements

All participants were examined in the Hanaro Medical Foundation (Seoul, Korea) or Dong-A Radiology Clinics (Daejeon, Korea) for adverse events and side effects. Safety monitoring, including vital signs (blood pressures, pulse rate, body temperature), was carried out on every visit. Laboratory tests were conducted at baseline and week 12 for hematologic (WBC, RBC, Hb, Hct, PLT, neutrophils, eosinophils, basophils, lymphocytes, monocytes), blood biochemical (AST, ALT, ALP, BUN, creatinine, e-GFR), and urine analysis (pH, nitrite, specific gravity, protein, glucose, ketone, bilirubin, blood, urobilinogen, color). The endometrial thickness was determined by transvaginal or abdominal ultrasonography at baseline and week 12 to monitor side effects. Adverse events were monitored throughout the study.

2.7. Statistical Analysis

Data analyses were performed using the intention-to-treat (ITT) analysis and tested for normal distribution graphically by evaluating quantile-quantile (QQ) plots. Values that exceed three times the interquartile range (IQR) (less than $Q1 - (3.0 \times IQR)$ or more than $Q3 + (3.0 \times IQR)$) were considered outliers and excluded from the analysis. Differences in the baseline characteristics between the two groups were tested with the Student's *t*-test for continuous variables and chi-squared or Fisher's exact test for categorical variables. Safety and effectiveness comparisons between or within groups were analyzed using a linear mixed-effects model with a random subject effect and fixed effects (group, week, group, the interaction between group and week) after adjustment for covariates. Covariate screening was analyzed using Empower (X&Y Solutions, Inc., Boston, MA, USA). Associations between HF scores or HF severity and bone turnover markers were tested in linear regression models. Data were presented as means and standard errors. All statistical analysis was performed using SAS (version 9.4; SAS Institute, Inc., Cary, NC, USA).

3. Results

3.1. Baseline Characteristics

Of the 147 subjects recruited, 84 eligible subjects were enrolled and randomized into either the KM or placebo group. Eight subjects dropped out before week 12 due to personal reasons ($n = 6$), taking an omega-3 supplement ($n = 1$), and participating in another clinical trial ($n = 1$). Finally, 76 participants (90%) completed the trial (Figure 1).

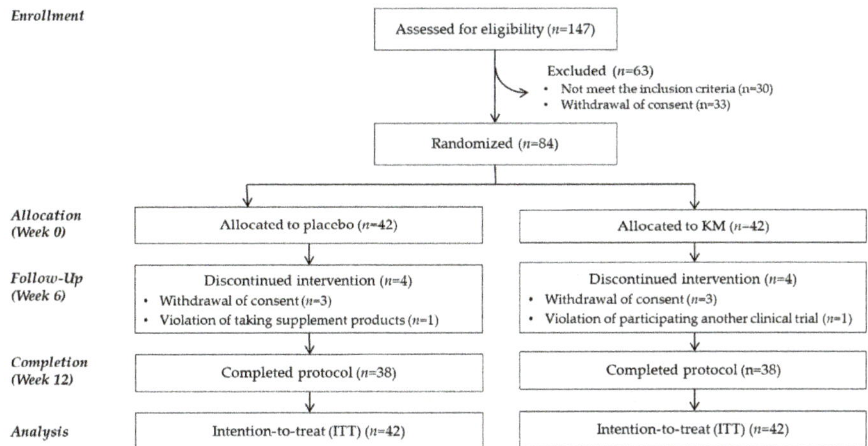

Figure 1. CONSORT flow diagram of the study. CONSORT, Consolidated Standards of Reporting Trials; KM, an extract mixture of kudzu flower and mandarin peel.

Table 1 shows the baseline characteristics of the subjects in ITT analysis. Study groups were well matched, with no significant differences between the KM and placebo groups, demonstrating that the participants were symptomatic peri- or post-menopausal women with a mean age of 51.8 ± 0.4 years, KI of 24.7, and HF score of 31.3. Although the difference in RFS between the two groups was statistically significant (27.0 vs. 23.0), they were all classified as the low fruit/vegetable consumption group [26]. The compliance was excellent in both arms (97.3% vs. 97.5%).

Table 1. Baseline characteristics of subjects in the intention-to-treat analysis.

Variables	Placebo (*n* = 42)	KM (*n* = 42)	*p*-Value
Age (years)	52.0 ± 0.6	51.5 ± 0.5	0.526
Amenorrhea period (months)	29.8 ± 4.5	20.7 ± 2.9	0.096
Hot flash score	34.8 ± 6.7	27.8 ±3.3	0.354
Kupperman Index	25.2 ± 1.3	24.1 ± 1.3	0.532
Alcohol drinker (Y/N)	21/21	19/23	0.662
Alcohol amount (g/week)	8.0 ± 2.2	6.5 ± 1.8	0.598
Smoker (Y/N)	0/42	1/41	1.000
Smoking amount (cigarettes/day)	0.0 ± 0.0	0.1 ± 0.1	0.183
Body weight (kg)	59.8 ± 1.5	57.7 ± 1.2	0.268
BMI (kg/m^2)	23.7 ± 0.5	23.1 ± 0.5	0.352
Waist circumference (cm)	83.0 ± 1.6	79.1 ± 1.3	0.059
Physical activity (MET-min/week)	1499.5 ± 148.7	1905.2 ± 251.7	0.170
Vigorous activity (h/week)	0.4 ± 0.2	0.2 ± 0.1	0.435
RFS	23.0 ± 1.2	27.0 ± 1.3	0.028
SBP (mmHg)	118.2 ± 1.6	118.2 ± 2.1	0.979
DBP (mmHg)	73.2 ± 1.4	72.7 ± 1.6	0.807

Values are presented as mean ± SE or *n*. Student's *t*-test for continuous variables and chi-square or Fisher's exact test for categorical variables were used to compare the difference between the treatments. KM, an extract mixture of kudzu flower and mandarin peel; BMI, body mass index; RFS, recommended food score; SBP, systolic blood pressure; DBP, diastolic blood pressure.

3.2. Effect of KM on HFs

Figure 2a–c illustrate the weekly changes in HF scores, severity, and frequency over the study period. These values were calculated based on Web-based digital diaries that record everyday changes. In both groups, all values were reduced substantially from the first week until the end of the study.

However, the overall reductions were lower in the KM group than in the placebo group, showing a statistical difference between the two groups in HF severity ($\beta = -0.38$, $p = 0.006$). As a result, at the end of the study, both the HF scores ($p = 0.041$) and HF severity ($p < 0.001$) were significantly lower in the KM group than in the placebo group. HF scores decreased by 60.1% in the KM group compared with 50.9% in the placebo group from the baseline, while HF severity decreased by 40% in the KM group compared with 26.3% in the placebo group from the baseline. Interestingly, the improvements in the HF scores and HF severity were better at night-time than at daytime, as visualized in the heat map (Figure 2d).

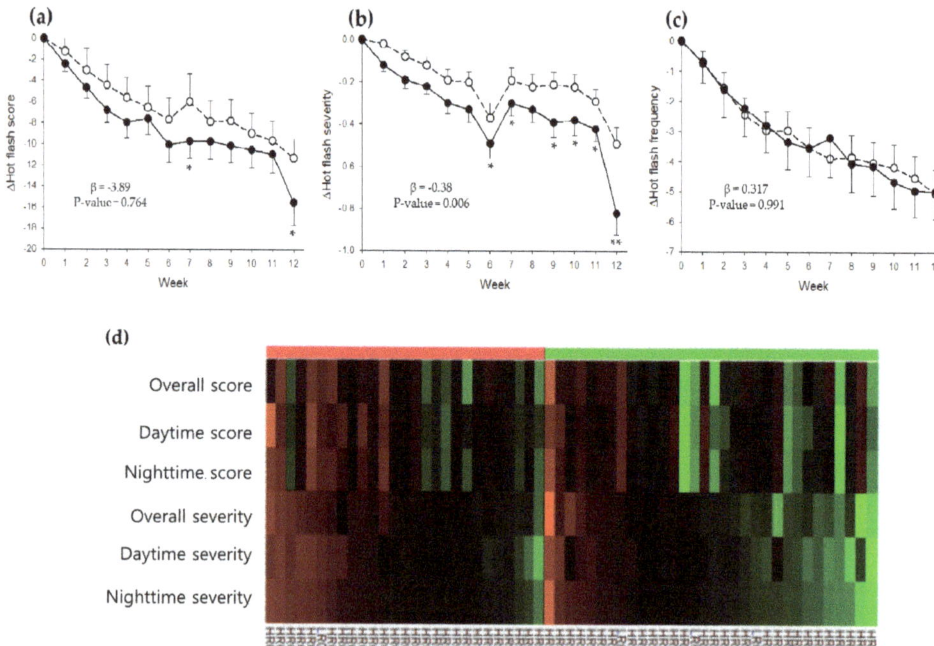

Figure 2. Changes in mean HF score (**a**), severity (**b**), and frequency (**c**) from baseline over the 12-week study period. Dots are experimental points (KM: black circles, placebo: empty circles). Heat map representing the changes in HF score and severity from baseline to the end of week 12 (**d**). The HF scores were calculated by multiplying the frequency experienced at that severity level. Then, the resulting points were summed to give a total daily score. Each daily score was averaged per week. Each value represents the mean + SE. Each column of the map represents one subject with all data for HF score and severity at total, daytime, and night-time in the heatmap. KM, an extract mixture of kudzu flower and mandarin peel; HF, hot flash. The *p*-values were obtained from a linear mixed-effect model adjusted for covariates. * $p < 0.05$, ** $p < 0.001$.

When using the KI and MenQOL retrospective questionnaires, the differences in HF scores' changes were not statistically significant between the placebo and KM groups. However, the within-group scores were significantly lower after the 12-week intervention than at baseline for both groups, indicating a trend for improving ($p < 0.05$ for all; upper part of Table 2). In the meantime, we determined various safety parameters, including vital signs and hematologic, blood biochemical, and urine analysis. The analysis showed that these values are normal, with no significant differences between the two groups at post-treatment (Supplementary Table S1). Besides, we found that there were no significant differences in endometrial thickness and LH level between the two groups. Although the FSH ($\beta = -10.81$, $p = 0.037$) and estradiol ($\beta = 46.23$, $p = 0.043$) levels were statistically different, all values remained normal ranges for peri- or post-menopausal women (lower part of Table 2). We note that, the FSH level rises above 40 mIU/mL during menopause [30], while estradiol level decreases from 30 to 400 pg/mL to 0–30 pg/mL [31]. We also note that the changes were opposite in the placebo and KM groups, indicating a potential protective capacity KM to suppress the hormonal changes in menopause. Finally, there were no significant adverse events and no difference in the total number of adverse events between them.

Table 2. Changes in retrospective HF scores and safety parameters.

Variables		Placebo ($n = 42$)	KM ($n = 42$)	Estimate	p-Value
		Retrospective HF scores			
KI	Week 0	6.86 ± 0.44	5.24 ± 0.37	0.83	0.191
	Week 12	4.63 ± 0.47	4.21 ± 0.4		
	p-value	<0.001	0.013		
MenQOL score	Week 0	15.17 ± 0.7	13.64 ± 0.64	0.555	0.163
	Week 12	9.97 ± 0.79	10.26 ± 0.75		
	p-value	<0.001	<0.001		
		Safety parameters			
Endometrial thickness (mm)	Week 0	5.4 ± 0.4	5.4 ± 0.6	0.52	0.380
	Week 12	5.0 ± 0.4	5.6 ± 0.5		
	p-value	0.14	0.793		
FSH (mIU/mL)	Week 0	63.4 ± 5.8	71.0 ± 6.5	−10.81	0.037
	Week 12	68.4 ± 5.5	66.2 ± 7.5		
	p-value	0.062	0.271		
Estradiol (pg/mL)	Week 0	42.1 ± 12.8	37.5 ± 11.6	46.23	0.043
	Week 12	23.6 ± 8.4	67.8 ± 19.1		
	p-value	0.192	0.115		
LH (mIU/mL)	Week 0	34.7 ± 2.8	35.9 ± 3.1	−0.09	0.986
	Week 12	38.3 ± 2.7	40.1 ± 4.5		
	p-value	0.27	0.267		

Values are presented as mean ± SE. Estimates and p-values were obtained from a linear mixed-effect model adjusted for covariates. KI, kupperman index; KM, an extract mixture of kudzu flower and mandarin peel; HF, hot flash; MenQOL, menopause quality of life scale; FSH, follicle stimulating hormone; LH, luteinizing hormone.

3.3. Effect of KM on Bone Turnover Markers

NTx and CTx, the amino- and carboxyterminal cross-linked telopeptides of type I collagen, are two bone resorption markers widely used in the clinical research setting [32]. In this study, the 12-week consumption of KM led to a significant and negative impact on CTx compared with the placebo ($\beta = -0.05$, $p = 0.027$). A similar result was obtained for NTx ($\beta = -0.80$), although neither reached statistical significance because of the high variability (upper part of Table 3).

Bone formation markers are synthesized by active osteoblasts and present in the circulation. OC is a hydroxyapatite-binding protein mainly synthesized by osteoblasts, while ALP is a membrane-bound glycoprotein present in four isoforms in the liver, intestine, placenta, and bone. BALP plays a role in osteoid formation and calcification by enzymatic degradation of pyrophosphate, a naturally occurring inhibitor of mineralization [33]. In this study, the KM group showed positive effects for both OC ($\beta = 1.03$) and BALP ($\beta = 0.67$) compared with the placebo group, although there were no significant statistical differences in the levels of OC and BALP between the two groups. Moreover, the difference in OC levels between baseline and the end of the study was significantly different after the 12-week consumption of KM ($p = 0.001$; lower part of Table 3).

Table 3. Changes in bone turnover markers in the blood.

Variables		Placebo ($n = 42$)	KM ($n = 42$)	Estimate	*p*-Value
		Markers of bone resorption			
CTx (ng/mL)	Week 0	0.43 ± 0.03	0.43 ± 0.03	−0.05	0.027
	Week 12	0.47 ± 0.03	0.39 ± 0.03		
	p-value	0.098	0.134		
NTx (ng/mL)	Week 0	3.50 ± 0.32	3.94 ± 0.45	−0.80	0.25
	Week 12	3.53 ± 0.36	3.06 ± 0.23		
	p-value	0.922	0.124		
		Markers of bone formation			
OC (mg/mL)	Week 0	18.35 ± 0.92	17.64 ± 0.98	1.03	0.17
	Week 12	19.49 ± 0.87	19.48 ± 1.12		
	p-value	0.124	0.001		
BALP (mg/mL)	Week 0	15.21 ± 0.73	13.48 ± 0.73	0.67	0.276
	Week 12	14.73 ± 0.70	13.53 ± 0.84		
	p-value	0.443	0.435		

Values are presented as mean ± SE. Estimate and p-value were obtained from a linear mixed-effect model adjusted for covariates. KM, an extract mixture of kudzu flower and mandarin peel; CTx, cross-linked C-telopeptides of bone collagen; NTx, cross-linked N-telopeptide of bone collagen; OC, osteocalcin; BALP, bone-specific alkaline phosphatase.

3.4. Associations between HF Scores and Bone Turnover Markers

Associations between HF scores and bone turnover markers were tested using multivariable linear regression. The results showed significant inverse correlations between the changes in HF score and OC ($r = -0.264$, $p = 0.031$), as well as those in HF severity and OC ($r = -0.228$, $p = 0.056$). In parallel with these results, the changes in CTx were positively correlated with those in HF score and HF severity, although they did not reach statistical significance (Figure 3).

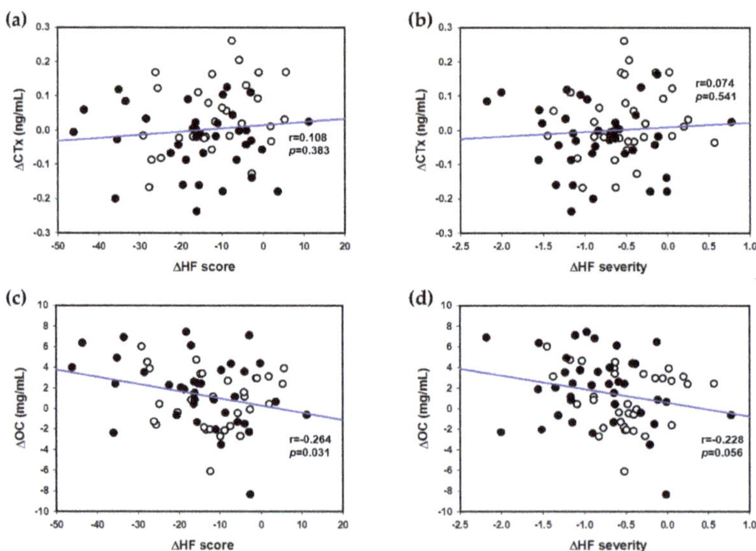

Figure 3. Associations between differential scores and bone turnover markers. CTx was positively correlated with HF scores (**a**) and HF severity (**b**), while OC was negatively correlated with HF scores (**c**) and HF severity (**d**). Dots are experimental points (KM: black circles, placebo: empty circles), and the continuous line indicates the straight-line fit by linear regression. KM, an extract mixture of kudzu flower and mandarin peel; HF, hot flashes; CTx, cross-linked C-telopeptides of bone collagen; OC, osteocalcin.

4. Discussion

The results of this randomized controlled clinical trial involving 84 peri- and post-menopausal women with moderate HFs support the hypothesis that KM might be an effective intervention for achieving reductions in HFs and improvements in bone turnover markers, as compared with the placebo. Another significant outcome of this clinical trial was the verified safety of KM. High-dose consumption of some isoflavones was shown to increase the risks of endometrial hyperplasia because of a high ER-binding affinity [34]. Therefore, we measured endometrial thickness and observed vital signs in all participants to provide precautionary measures and suggestions. As a result, we confirmed no difference in the incidence of adverse events between the two groups. These data provide additional support for the safety of KM use [27,28]. Kim et al. [35] presented an in vitro study highlighting a potential role of kudzu flower extract as an anti-endometriotic agent with inhibition against endometriotic cell adhesion and migration. In a mouse model of menopause, Sternlicht et al. [36] showed that kudzu flower extract exerted a noticeable down-regulation of matrix metalloproteinase, which regulates migration, invasion, and proliferation of various cells.

Although not fully understood, estrogen withdrawal is thought to trigger several neuroendocrine pathways in humans, ultimately causing thermoregulatory dysfunction in the hypothalamus. As the thermoregulatory zone becomes lower and narrower in menopausal women, a small change in core body temperature may trigger the heat loss mechanisms, leading to HF symptoms [37]. In the present study, the KM and placebo groups showed a 60.1% and 50.9% reduction in HF score and a 40.0% and 26.3% reduction in HF severity, respectively, from baseline to the end of treatment. These rates of reduction in HF score and severity are comparable or better than those achieved through non-hormonal interventions, including soy isoflavones, black cohosh, and red clover [38]. It is reported that a mixture of tectorigenin derivatives dramatically reduces the OVX-induced rise in tail skin temperature in mice [17]. Tectorigenin derivatives are unique isoflavones found in kudzu flower but not in soy [39]. Hence, KM's potential as an effective non-hormonal alternative for the management of menopausal HFs might be attributed to the presence of tectorigenin derivatives. Additionally, we extended our findings to compare the impact of KM on the day and night HF outcomes using a heat map. Results showed a better improvement in night-time HFs than daytime HFs. This observation suggests that KM might also improve sleep quality in individuals with night-time HFs but needs further study.

The measurement of HFs can present unique methodologic challenges due to intangible and subjective characteristics [27]. Traditional retrospective summative HF diaries are inaccurate measures of actual HF status because of the subject's memory capacity limitations and the influence of mood on symptom reporting [28]. We thus decided to use prospective electronic digital diaries over 12 weeks from 7 days before treatment. This approach is not only a cost-effective data acquisition scheme but also relatively more precise in quantifying HFs because it reduces the limitations of recall [28]. Furthermore, the current study supplemented the severity and frequency data of menopausal symptoms with the self-administered questionnaires to monitor quality-of-life as a treatment outcome. Although the results demonstrated similar improvements, there were no significant differences between the two groups. The failure of statistical significance is likely due to both the less frequent measurement intervals and the higher dependence on subjective perception compared with the electronic digital diaries. Moreover, although validated, the KI and MenQOL have weaknesses, such as limited focus and lack of psychometric assessments [40].

The onset of menopausal HFs is associated with various adverse disease risk factors [41]. In the current study, we explored the associations between HFs and bone turnover markers because estrogen deficiency in menopause is related to rapid bone loss [42]. Notably, the changes in HFs were positively correlated with those in CTx and negatively correlated with those in OC. That is, there seemed to be higher bone resorption and lower bone formation among subjects with more HFs. These results were supported by earlier animal studies where KM, tectorigenin derivatives, and hesperidin were all effective in down-regulating receptor activator of nuclear factor κB ligand (RANKL) in OVX mice [17,18]. RANKL is a transmembrane ligand expressed on osteoblasts, which activates RANK in osteoclasts and triggers osteoclast maturation and bone resorption [43]. Other studies also indicated the inhibitory activities of tectorigenin derivatives against RANKL-induced osteoclastogenesis [44,45], and a preventive role of hesperidin against bone loss in OVX animals [16,46]. However, mixed results have been reported regarding the association between HFs and BMD in humans. Salamone et al. [47] observed a noticeably lowered BMD in the spine, hip, and whole body in 290 women reporting HFs. Similarly, Lee and Kanis [48] showed a positive association between VMS and vertebral fractures. In contrast, Scoutellas et al. [49] found no association between VMS and vertebral fractures in post-menopausal women aged 50–64 years. It was suggested that these conflicting results might originate from differences in subject ages, use of hormone therapy, or recall period [50].

Limitations of this study include that we did not measure levels of serum serotonin and epinephrine, which are thought to play an important role in thermoregulatory function. We also did not measure the levels of RANKL gene expression responsible for the bone turnover function. As a result, direct mechanisms of action of KM in relieving HFs and bone turnover were not fully understood. Nonetheless, this randomized controlled trial provides a novel finding that KM appears to alleviate HFs during the menopausal transition. The KM was well-tolerated in this study. Besides, this study had many strengths. First, we included peri- and post-menopausal women because of the peak in HF severity during the latter part of the menopausal transition near the final menstrual period [51]. Second, HF data were collected prospectively using electronic digital diaries, and daytime and night-time HFs were also analyzed separately. Third, the intervention lasted for 12 weeks, which is a relatively longer follow-up than most of the other studies [52]. Lastly, KM has been authenticated by chemical analysis of three tectorigenin derivatives and hesperidin.

5. Conclusions

In conclusion, given the positive effects of KM on HFs and bone turnover and the favorable side effect profile, supplementation with a KM dose of 1150 mg/day seems to be an acceptable option for reducing HF symptoms, as well as improving bone turnover, compared with the placebo group, during the menopausal transition. Future studies are needed to investigate the underlying mechanisms behind the effects and synergistic interactions between tectorigenin derivatives and hesperidin on HFs and bone turnover.

Supplementary Materials: The following are available online at http://www.mdpi.com/2072-6643/12/11/3237/s1, Table S1: Safety parameters.

Author Contributions: Conceptualization, O.K. and J.L.; methodology, O.K.; formal analysis, J.E.K., H.J. and S.H.; investigation, J.E.K. and H.J.; resources, J.L.; data curation, J.E.K., H.J. and O.K.; writing—original draft preparation, J.E.K.; writing—review and editing, O.K.; visualization, J.E.K. and H.J.; supervision, O.K.; project administration, H.J.; funding acquisition, O.K. All authors have read and agreed to the published version of the manuscript.

Funding: This research was funded by the Bio-Synergy Research Project of the Ministry of Science and ICT through the National Research Foundation, grant number NRF-2012M3A9C4048761, and the LG Household and Healthcare.

Acknowledgments: We thank the 84 women who participated in this study, and Hanaro Medical Foundation, and Dong-A Radiology Clinics for contributing to the safety monitoring over the entire study period. We also thank our study physicians, Bumjo Oh and KyuRi Hwang, for their clinical oversight of the women in this study.

Conflicts of Interest: The authors declare no conflict of interest. J.L. is affiliated with LG Household and Healthcare. The funders had no role in the study design; in the collection, analyses, or interpretation of data; in the writing of the manuscript, or in the decision to publish the results.

References

1. Dugan, S.A.; Powell, L.H.; Kravitz, H.M.; Everson Rose, S.A.; Karavolos, K.; Luborsky, J. Musculoskeletal pain and menopausal status. *Clin. J. Pain* **2006**, *22*, 325–331. [CrossRef] [PubMed]

2. Vousoura, E.; Spyropoulou, A.C.; Koundi, K.L.; Tzavara, C.; Verdeli, H.; Paparrigopoulos, T.; Augoulea, A.; Lambrinoudaki, I.; Zervas, I.M. Vasomotor and depression symptoms may be associated with different sleep disturbance patterns in postmenopausal women. *Menopause* **2015**, *22*, 1053–1057. [CrossRef] [PubMed]

3. Utian, W.H. Psychosocial and socioeconomic burden of vasomotor symptoms in menopause: A comprehensive review. *Health Qual. Life Outcomes* **2005**, *3*, 47. [CrossRef] [PubMed]

4. Nelson, H.D. Menopause. *Lancet* **2008**, *371*, 760–770. [CrossRef]

5. Landgren, B.M.; Collins, A.; Csemiczky, G.; Burger, H.G.; Baksheev, L.; Robertson, D.M. Menopause transition: Annual changes in serum hormonal patterns over the menstrual cycle in women during a nine-year period prior to menopause. *J. Clin. Endocr. Metab.* **2004**, *89*, 2763–2769. [CrossRef]

6. Deecher, D.C.; Dorries, K. Understanding the pathophysiology of vasomotor symptoms (hot flushes and night sweats) that occur in perimenopause, menopause, and postmenopause life stages. *Arch. Women's Ment. Health* **2007**, *10*, 247–257. [CrossRef]

7. Rodstrom, K.; Bengtsson, C.; Lissner, L.; Milsom, I.; Sundh, V.; Bjorkelund, C. A longitudinal study of the treatment of hot flushes: The population study of women in Gothenburg during a quarter of a century. *Menopause J. N. Am. Menopause Soc.* **2002**, *9*, 156–161. [CrossRef]

8. Abdia, F.; Alimoradi, Z.; Haqi, P.; Mandizad, F. Effects of phytoestrogens on bone mineral density during the menopause transition: A systematic review of randomized, controlled trials. *Climacteric* **2016**, *19*, 535–545. [CrossRef]

9. Crandall, C.J.; Zheng, Y.; Crawford, S.L.; Thurston, R.C.; Gold, E.B.; Johnston, J.M.; Greendale, G.A. Presence of vasomotor symptoms is mineral density: A longitudinal associated with lower bone analysis. *Menopause J. N. Am. Menopause Soc.* **2009**, *16*, 239–246. [CrossRef]

10. Tuomikoski, P.; Ylikorkala, O.; Mikkola, T.S. Postmenopausal hot flushes and bone mineral density: A longitudinal study. *Acta Obstet. Gyn. Scan.* **2015**, *94*, 198–203. [CrossRef]

11. Rosendaal, F.R.; Helmerhorst, F.M.; Vandenbroucke, J.P. Female hormones and thrombosis. *Arter. Thromb. Vasc. Biol.* **2002**, *22*, 201–210. [CrossRef] [PubMed]

12. Isoflavones Market by Source (Soy, Red Clover), Application (Pharmaceuticals, Nutraceuticals, Cosmetics, and Food & Beverages), Form (Powder and Liquid), and Region (North America, Europe, Asia Pacific, RoW)–Global Forecast to 2025. Available online: https://www.marketsandmarkets.com/Market-Reports/isoflavones-market-161667005.html (accessed on 22 October 2020).

13. Wang, S.; Gong, T.; Lu, J.; Kano, Y.; Yuan, D. Simultaneous determination of tectorigenin and its metabolites in rat plasma by ultra performance liquid chromatography/quadrupole time-of-flight mass spectrometry. *J. Chromatogr. B Analyt. Technol. Biomed. Life Sci.* **2013**, *933*, 50–58. [CrossRef] [PubMed]

14. Branka Levaj, V.D.-U.; Bursackovacevic, D.; Krasnici, N. Determination of flavonoids in pulp and peel of mandarin fruits. *Agric. Conspec. Sci.* **2009**, *74*, 221–225.

15. Lim, D.W.; Lee, Y.; Kim, Y.T. Preventive effects of Citrus unshiu peel extracts on bone and lipid metabolism in OVX rats. *Molecules* **2014**, *19*, 783–794. [CrossRef]

16. Chiba, H.; Uehara, M.; Wu, J.; Wang, X.; Masuyama, R.; Suzuki, K.; Kanazawa, K.; Ishimi, Y. Hesperidin, a citrus flavonoid, inhibits bone loss and decreases serum and hepatic lipids in ovariectomized mice. *J. Nutr.* **2003**, *133*, 1892–1897. [CrossRef] [PubMed]

17. Han, N.R.; Nam, S.Y.; Hong, S.; Kim, H.Y.; Moon, P.D.; Kim, H.J.; Cho, H.; Lee, B.; Kim, H.M.; Jeong, H.J. Improvement effects of a mixed extract of flowers of Pueraria thomsonii Benth. and peels of Citrus unshiu Markovich on postmenopausal symptoms of ovariectomized mice. *Biomed. Pharm.* **2018**, *103*, 524–530. [CrossRef] [PubMed]

18. Jeong, H.J.; Kim, M.H.; Kim, H.; Kim, H.Y.; Nam, S.Y.; Han, N.R.; Lee, B.; Cho, H.; Moon, P.D.; Kim, H.M. PCE17 and its active compounds exert an anti-osteoporotic effect through the regulation of receptor activator of nuclear factor-κB ligand in ovariectomized mice. *J. Food Biochem.* **2018**, *42*, e12561. [CrossRef]

19. Fabricant, C.H.A.D. *Pre-DSHEA List of Old Dietary Ingredients*; Natural Products Association (NPA): Washington, DC, USA, 2017.

20. Chang, A.; Kwak, B.Y.; Yi, K.; Kim, J.S. The effect of herbal extract (EstroG-100) on pre-, peri- and post-menopausal women: A randomized double-blind, placebo-controlled study. *Phytother. Res.* **2012**, *26*, 510–516. [CrossRef]

21. Kim, S.Y.; Seo, S.K.; Choi, Y.M.; Jeon, Y.E.; Lim, K.J.; Cho, S.; Choi, Y.S.; Lee, B.S. Effects of red ginseng supplementation on menopausal symptoms and cardiovascular risk factors in postmenopausal women: A double-blind randomized controlled trial. *Menopause* **2012**, *19*, 461–466. [CrossRef]

22. Zhou, X.; Han, D.; Xu, R.; Li, S.; Wu, H.; Qu, C.; Wang, F.; Wang, X.; Zhao, Y. A model of metabolic syndrome and related diseases with intestinal endotoxemia in rats fed a high fat and high sucrose diet. *PLoS ONE* **2014**, *9*, e115148. [CrossRef]

23. Li, M.D.; Hung, A.; Lenon, G.B.; Yang, A.W.H. Chinese herbal formulae for the treatment of menopausal hot flushes: A systematic review and meta-analysis. *PLoS ONE* **2019**, *14*, e0222383. [CrossRef]

24. Committee, I.R. Guidelines for Data Processing and Analysis of the International Physical Activity Questionnaire (IPAQ)-Short and Long Forms. 2005. Available online: http://www.ipaq.ki.se/scoring.pdf (accessed on 22 October 2020).

25. Buysse, D.J.; Reynolds, C.F., 3rd; Monk, T.H.; Berman, S.R.; Kupfer, D.J. The Pittsburgh Sleep Quality Index: A new instrument for psychiatric practice and research. *Psychiatry Res.* **1989**, *28*, 193–213. [CrossRef]

26. Kim, J.Y.; Yang, Y.J.; Yang, Y.K.; Oh, S.Y.; Hong, Y.C.; Lee, E.K.; Kwon, O. Diet quality scores and oxidative stress in Korean adults. *Eur. J. Clin. Nutr.* **2011**, *65*, 1271–1278. [CrossRef] [PubMed]

27. Sloan, J.A.; Loprinzi, C.L.; Novotny, P.J.; Barton, D.L.; Lavasseur, B.I.; Windschitl, H. Methodologic lessons learned from hot flash studies. *J. Clin. Oncol.* **2001**, *19*, 4280–4290. [CrossRef]

28. Fisher, W.I.; Thurston, R.C. Measuring hot flash phenomenonology using ambulatory prospective digital diaries. *Menopause* **2016**, *23*, 1222–1227. [CrossRef]

29. Hilditch, J.R.; Lewis, J.; Peter, A.; van Maris, B.; Ross, A.; Franssen, E.; Guyatt, G.H.; Norton, P.G.; Dunn, E. A menopause-specific quality of life questionnaire: Development and psychometric properties. *Maturitas* **1996**, *24*, 161–175. [CrossRef]

30. Zhu, D.; Li, X.; Macrae, V.E.; Simoncini, T.; Fu, X. Extragonadal Effects of Follicle-Stimulating Hormone on Osteoporosis and Cardiovascular Disease in Women during Menopausal Transition. *Trends Endocrinol. Metab.* **2018**, *29*, 571–580. [CrossRef]

31. Lee, J.S.; Ettinger, B.; Stanczyk, F.Z.; Vittinghoff, E.; Hanes, V.; Cauley, J.A.; Chandler, W.; Settlage, J.; Beattie, M.S.; Folkerd, E.; et al. Comparison of methods to measure low serum estradiol levels in postmenopausal women. *J. Clin. Endocrinol. Metab.* **2006**, *91*, 3791–3797. [CrossRef]

32. Herrmann, M.; Seibel, M.J.; Seibel, M. The amino- and carboxyterminal cross-linked telopeptides of collagen type I, NTX-I and CTX-I: A comparative review. *Clin. Chim. Acta* **2008**, *393*, 57–75. [CrossRef] [PubMed]

33. Shetty, S.; Kapoor, N.; Bondu, J.D.; Thomas, N.; Paul, T.V. Bone turnover markers: Emerging tool in the management of osteoporosis. *Indian J. Endocrinol. Metab.* **2016**, *20*, 846–852. [CrossRef]

34. Lu, L.J.W.; Anderson, K.E.; Grady, J.J.; Kohen, F.; Nagamani, M. Decreased ovarian hormones during a soya diet: Implications for breast cancer prevention. *Cancer Res.* **2000**, *60*, 4112–4121. [PubMed]

35. Kim, J.H.; Woo, J.H.; Kim, H.M.; Oh, M.S.; Jang, D.S.; Choi, J.H. Anti-Endometriotic Effects of Pueraria Flower Extract in Human Endometriotic Cells and Mice. *Nutrients* **2017**, *9*, 212. [CrossRef] [PubMed]

36. Sternlicht, M.D.; Werb, Z. How matrix metalloproteinases regulate cell behavior. *Annu. Rev. Cell Dev. Biol.* **2001**, *17*, 463–516. [CrossRef] [PubMed]

37. Perez, D.G.; Loprinzi, C.L.; Barton, D.L.; Pockaj, B.A.; Sloan, J.; Novotny, P.J.; Christensen, B.J. Pilot evaluation of mirtazapine for the treatment of hot flashes. *J. Support Oncol.* **2004**, *2*, 50–56.

38. Carroll, D.G. Nonhormonal therapies for hot flashes in menopause. *Am. Fam. Physician* **2006**, *73*, 457–464.

39. Kamiya, T.; Nagamine, R.; Sameshima-Kamiya, M.; Tsubata, M.; Ikeguchi, M.; Takagaki, K. The isoflavone-rich fraction of the crude extract of the Puerariae flower increases oxygen consumption and BAT UCP1 expression in high-fat diet-fed mice. *Glob. J. Health Sci.* **2012**, *4*, 147–155. [CrossRef]

40. Simon, J.A.; Kaunitz, A.M.; Kroll, R.; Graham, S.; Bernick, B.; Mirkin, S. Oral 17β-estradiol/progesterone (TX-001HR) and quality of life in postmenopausal women with vasomotor symptoms. *Menopause* **2019**, *26*, 506–512. [CrossRef]

41. Thurston, R.C.; El Khoudary, S.R.; Sutton-Tyrrell, K.; Crandall, C.J.; Gold, E.B.; Sternfeld, B.; Joffe, H.; Selzer, F.; Matthews, K.A. Vasomotor symptoms and lipid profiles in women transitioning through menopause. *Obstet. Gynecol.* **2012**, *119*, 753–761. [CrossRef]

42. Massé, P.G.; Dosy, J.; Jougleux, J.L.; Caissie, M.; Howell, D.S. Bone mineral density and metabolism at an early stage of menopause when estrogen and calcium supplement are not used and without the interference of major confounding variables. *J. Am. Coll. Nutr.* **2005**, *24*, 354–360. [CrossRef]

43. Bezerra, M.C.; Carvalho, J.F.; Prokopowitsch, A.S.; Pereira, R.M. RANK, RANKL and osteoprotegerin in arthritic bone loss. *Braz. J. Med. Biol. Res.* **2005**, *38*, 161–170. [CrossRef]

44. Ma, C.Y.; Xu, K.; Meng, J.H.; Ran, J.S.; Moqbel, S.A.A.; Liu, A.; Yan, S.G.; Wu, L.D. Tectorigenin inhibits RANKL-induced osteoclastogenesis via suppression of NF-kappa B signalling and decreases bone loss in ovariectomized C57BL/6. *J. Cell. Mol. Med.* **2018**, *22*, 5121–5131. [CrossRef]

45. Wang, J.; Tang, Y.; Lv, X.; Zhang, J.; Ma, B.; Wen, X.; Bao, Y.; Wang, G. Tectoridin inhibits osteoclastogenesis and bone loss in a murine model of ovariectomy-induced osteoporosis. *Exp. Gerontol.* **2020**, 111057. [CrossRef] [PubMed]

46. Horcajada, M.N.; Habauzit, V.; Trzeciakiewicz, A.; Morand, C.; Gil-Izquierdo, A.; Mardon, J.; Lebecque, P.; Davicco, M.J.; Chee, W.S.S.; Coxam, V.; et al. Hesperidin inhibits ovariectomized-induced osteopenia and shows differential effects on bone mass and strength in young and adult intact rats. *J. Appl. Physiol.* **2008**, *104*, 648–654. [CrossRef] [PubMed]

47. Salamone, L.M.; Gregg, E.; Wolf, R.L.; Epstein, R.S.; Black, D.; Palermo, L.; Kuller, L.H.; Cauley, J.A. Are menopausal symptoms associated with bone mineral density and changes in bone mineral density in premenopausal women? *Maturitas* **1998**, *29*, 179–187. [CrossRef]

48. Lee, S.J.; Kanis, J.A. An association between osteoporosis and premenstrual symptoms and postmenopausal symptoms. *Bone Miner.* **1994**, *24*, 127–134. [CrossRef]

49. Scoutellas, V.; O'Neill, T.W.; Lunt, M.; Reeve, J.; Silman, A.J. Does the presence of postmenopausal symptoms influence susceptibility to vertebral deformity? European Vertebral Osteoporosis Study (EVOS) Group. *Maturitas* **1999**, *32*, 179–187. [CrossRef]

50. Pal, L.; Bevilacqua, K.; Zeitlian, G.; Shu, J.; Santoro, N. Implications of diminished ovarian reserve (DOR) extend well beyond reproductive concerns. *Menopause* **2008**, *15*, 1086–1094. [CrossRef]

51. Woods, N.F.; Mitchell, E.S. Symptoms during the perimenopause: Prevalence, severity, trajectory, and significance in women's lives. *Am. J. Med.* **2005**, *118* (Suppl. 2), 14–24. [CrossRef]

52. Shams, T.; Firwana, B.; Habib, F.; Alshahrani, A.; Alnouh, B.; Murad, M.H.; Ferwana, M. SSRIs for hot flashes: A systematic review and meta-analysis of randomized trials. *J. Gen. Intern. Med.* **2014**, *29*, 204–213. [CrossRef]

Publisher's Note: MDPI stays neutral with regard to jurisdictional claims in published maps and institutional affiliations.

© 2020 by the authors. Licensee MDPI, Basel, Switzerland. This article is an open access article distributed under the terms and conditions of the Creative Commons Attribution (CC BY) license (http://creativecommons.org/licenses/by/4.0/).

Article

Depressive Symptoms in Middle-Aged and Elderly Women Are Associated with a Low Intake of Vitamin B6: A Cross-Sectional Study

Tamami Odai [1], Masakazu Terauchi [1,*], Risa Suzuki [2], Kiyoko Kato [1], Asuka Hirose [2] and Naoyuki Miyasaka [2]

[1] Department of Women's Health, Tokyo Medical and Dental University, Tokyo 113-8510, Japan; odycrm@tmd.ac.jp (T.O.); kiyo.crm@tmd.ac.jp (K.K.)
[2] Department of Obstetrics and Gynecology, Tokyo Medical and Dental University, Tokyo 113-8510, Japan; 150612ms@tmd.ac.jp (R.S.); a-kacrm@tmd.ac.jp (A.H.); n.miyasaka.gyne@tmd.ac.jp (N.M.)
* Correspondence: teragyne@tmd.ac.jp; Tel./Fax: +81-3-5803-4605

Received: 6 October 2020; Accepted: 4 November 2020; Published: 9 November 2020

Abstract: This study investigated the nutritional factors that are associated with anxiety and depressive symptoms in Japanese middle-aged and elderly women. We conducted a cross-sectional study with 289 study participants aged $\geq$40 years (mean age = 52.0 $\pm$ 6.9 years). Their dietary habits, menopausal status and symptoms, and varied background factors, such as body composition, lifestyle factors, and cardiovascular parameters, were assessed. Their anxiety and depressive symptoms were evaluated using the Hospital Anxiety and Depression Scale (HADS), where scores of 0–7 points, 8–10 points, and 11–21 points on either the anxiety or depression subscales were categorized as mild, moderate, and severe, respectively. The dietary consumption of nutrients was assessed using a brief self-administered diet history questionnaire. The relationships between the moderate-to-severe anxiety/depressive symptoms and the dietary intake of 43 major nutrients were investigated using multivariate logistic regression analyses. After adjusting for age, menopausal status, and the background factors that were significantly related to depressive symptoms, moderate and severe depression was significantly inversely associated with only vitamin B6 (adjusted odds ratio per 10 µg/MJ in vitamin B6 intake = 0.89, 95% confidence interval = 0.80–0.99). A higher intake of vitamin B6 could help relieve depressive symptoms for this population.

Keywords: depression; anxiety; menopause; mental disorder; pyridoxal 5′-phosphate

1. Introduction

Depression and anxiety are common mental disorders. According to the World Health Organization, the estimated number of people globally who have depression and anxiety has increased by 18.4% and 14.9%, respectively, over the last ten years, totaling more than 300 million and 250 million, respectively [1]. Women are more likely to have mental disorders than men [1]. Large fluctuations of the serum estrogen level in the premenstrual, postpartum, and perimenopausal periods contribute to mood changes [2,3]. An increased risk for mental disorders, including depression, is observed during the menopause transition [4,5]; additionally, the early menopause transition and early menopause are significant risks for depression, which indicates that a short duration of exposure to endogenous estrogens can increase the risk for late-life depression [6,7]. Estrogen plays neuroprotective, anti-depressive, and anti-anxiety roles via the regulation of serotonergic and noradrenergic systems [8]; however, the associations between psychological symptoms and absolute serum sex hormone levels are inconsistent [3,9]. Additionally, several reports have demonstrated the beneficial effects of hormone replacement therapy (HRT) on mood symptoms [10–12], while a randomized controlled trial covering a wide age range of postmenopausal women failed to show the effects of HRT on depressive disorders [13].

Furthermore, as there are increasing concerns about the side effects of HRT associated with cardiovascular diseases and hormone-sensitive cancers, such as breast, endometrial, and ovarian cancer, expectations for a complementary form of therapy are growing. Several reports have shown the association between specific nutrients, such as thiamine, folate, vitamin B6 and B12, zinc, and iron, and mental health [14,15]; however, the psychological effects of dietary intake of various nutrients remain largely unknown, especially in peri- and postmenopausal women. This study thus investigated the associations between anxiety/depressive symptoms and dietary consumption of nutrients in Japanese middle-aged and elderly women.

2. Materials and Methods

2.1. Study Population

This cross-sectional analysis of 700 Japanese women who had enrolled in the Systematic Health and Nutrition Education Program at the menopause clinic of the Tokyo Medical and Dental University was conducted based on the first-visit medical records from January 2009 to August 2017. All participants who registered in this program had sought medical attention for menopausal symptoms and had been provided improvement strategies based on the assessment of their physical and psychological health status and lifestyle.

We evaluated women's anxiety/depressive symptoms using the Hospital Anxiety and Depression Scale (HADS) and their dietary habits using a brief self-administered diet history questionnaire (BDHQ). A total of 313 participants who had failed to complete the HADS and/or BDHQ, 46 participants who had been treated with estrogen, four participants who were aged <40 years, and 48 participants whose menopausal status was uncertain were excluded. For the remaining 289 women, the associations between anxiety/depressive symptoms and the dietary intake of nutrients were evaluated.

This study protocol was reviewed and approved by the Tokyo Medical and Dental University Review Board (approval number: 774) and written informed consent was obtained from all participants. The research was performed in accordance with the Declaration of Helsinki.

2.2. Measurements

2.2.1. Menopausal Definitions

The women were defined as premenopausal if they had regular menstrual cycles, in the menopausal transition if they had menstruated within the past 12 months but had missed periods or had an irregular cycle in the past 3 months, and postmenopausal if they had not menstruated in the past 12 months. The postmenopausal women with surgically induced menopause via a hysterectomy and/or oophorectomy were excluded as their menopausal status was unclear.

2.2.2. Physical Assessment

Participants' body composition variables, including height, weight, body mass index, body fat percentage, fat mass, lean body mass, muscle mass, water mass, and basal metabolic rate, were measured using a bioimpedance analyzer (MC190-EM; Tanita, Tokyo, Japan). Waist and hip circumferences were measured to calculate the waist-to-hip ratio, and body temperature was measured with a thermometer. Resting energy expenditure was also determined based on the respiratory volume using an indirect calorimeter (Metavine-N VMB-005 N; Vine, Tokyo, Japan). Moreover, cardiovascular parameters, including systolic and diastolic blood pressure, heart rate, cardio-ankle vascular index, and ankle-brachial pressure index, were evaluated using a vascular screening system (VS-1000; Fukuda Denshi, Tokyo, Japan). Additionally, we utilized a physical fitness test to assess power, reaction time, and flexibility. Hand-grip strength was measured twice for each hand with a hand dynamometer (Yagami, Nagoya, Japan) and the mean value (kgf) was calculated using the larger value of the two measurements. The ruler drop test, which is an assessment of reaction time, was repeated seven times using a wooden ruler that was 60 cm in length and 110 g (Yagami, Nagoya, Japan). Participants fixed their arms on a desk and outstretched their fingers from the edge of the desk, while the bottom of the ruler was hung from between the thumb and index finger of an examiner; the participants then attempted to catch the ruler as quickly as possible when it was dropped. Where the participants gripped the ruler was evaluated, and the average reaction time (cm) was calculated from the remaining five values after the largest and smallest values were omitted. The sit-and-reach test for assessing flexibility was conducted with a reach box while sitting (Yagami, Nagoya, Japan).

2.2.3. Lifestyle Factors

We investigated lifestyle characteristics, such as the frequency of smoking (none, fewer than 20 cigarettes per day, or 20 or more cigarettes per day), consumption of alcohol (never, sometimes, or daily), and caffeinated drinks (none, once or twice per day, or three or more times per day), and regular exercise habits (yes or no).

2.2.4. Questionnaires

The HADS, which is a widely used screening instrument for anxiety and depression, was developed to evaluate the mental health of patients with somatic symptoms [16]. It consists of seven items across two subscales: anxiety and depression. Anxiety is assessed using feeling states: feeling tense, restless, or panicky; feeling something awful will happen; having worrying thoughts; feeling unable to relax; having butterflies in one's stomach. Furthermore, depression is evaluated using anhedonia: unable to enjoy things, unable to laugh and see the funny side, not feeling cheerful, feeling slowed down, having lost interest in one's appearance, unable to look forward to things, and unable to enjoy a book or TV. Each item is scored on a four-point Likert scale. Cut-off points to identify doubtful and definite cases for anxiety or depression are 8 and 11 points, respectively. In the present study, scores of 0–7 points, 8–10 points, and 11–21 points were classified as mild, moderate, and severe anxiety/depression, respectively.

The BDHQ, a short version of a self-administered diet history questionnaire that was developed in Japan, is composed of 77 questions and takes approximately 15 min to answer. The BDHQ was used to assess the intake frequency of 61 food items that are commonly consumed in Japan, mainly from the food list used in the National Health and Nutrition Survey of Japan [17], including beverages and seasonings, in the preceding month. Using an ad hoc computer algorithm for estimating the daily intake of nutrients and food after an adjustment for total calorie intake, the consumption of 96 nutrients and 58 food items was calculated. The estimated intake of nutrients and food items based on the BDHQ has previously been validated via a comparison with dietary records using a semi-weighted method [18,19]. In this study, we investigated the association between anxiety and depressive symptoms and 43 major nutrients with high validity (Table S1).

We also evaluated the women's health using the Menopausal Health-Related Quality of Life Questionnaire (MHR-QOL). The MHR-QOL, which is a modification of the Women's Health Questionnaire and others [20–22], was developed and validated in our clinic [23–26] and comprises four categories: physical health, psychological health, life satisfaction, and social involvement (Table S2). The items for physical and psychological symptoms are scored using a four-point Likert scale based on the symptom frequencies. According to the degree of agreement or disagreement, life satisfaction and social involvement were assessed using a two- or four-point Likert scale. We pooled the scores for somatic symptoms (6 items), vasomotor symptoms (2 items), insomnia symptoms (2 items), life satisfaction (5 items), and social involvement (12 items), to evaluate the score for each subcategory. A low score represented severe symptoms and low levels of life satisfaction and social engagement.

Lastly, we assessed psychotropic medication use, including hypnotics, anxiolytics, and antidepressants, based on medical interview responses.

2.3. Statistical Analysis

Continuous variables are presented as mean ± standard deviation. The required sample size was calculated as follows: 10 times the number of independent variables divided by the prevalence of anxiety or depressive symptoms, which were estimated to be 8 and 0.4, respectively, was 200. The comparison between groups was performed using the Kruskal–Wallis and the chi-squared tests. Cut-off points for the identification of multicollinearity were determined using a Pearson or Spearman correlation coefficient of $r > 0.9$. The nutrients and background factors associated with anxiety or depressive symptoms were evaluated using a stepwise regression procedure with a threshold of $p = 0.1$ and $p = 0.05$ for variable inclusion and exclusion, respectively. The analysis was conducted with a multivariate logistic regression model to clarify the relationships between the severity of anxiety and depressive symptoms and the selected nutrients. Significance was set at $p < 0.05$. All analyses were performed with GraphPad Prism version 5.02 (GraphPad Software, San Diego, CA, USA) and JMP version 12 (SAS Institute Inc, Cary, NC, USA).

3. Results

The participants' ($n = 289$) mean age was 52.0 ± 6.9 years. The prevalence of mild, moderate, and severe anxiety was 54.0%, 28.7%, and 17.3%, respectively, and that of depression was 61.6%, 25.6%, and 12.8%, respectively. The participants' background characteristics are shown in Tables 1 and 2. The participants with severe anxiety/depressive symptoms had a low quality of life according to the assessment using the MHR-QOL and were less frequently engaged in exercise. Moreover, the participants with severe depressive symptoms were younger than those with mild symptoms. There was no significant difference in body composition or the physical fitness test between the three severity groups.

First, we assessed the daily intake of 43 nutrients; then, we investigated the nutritional intake, which differed significantly between the three anxiety/depression severity groups. The intake of 10 and 22 nutrients showed significant differences between the three groups of anxiety and depression severity, respectively (Table 3 and Table S3). Similarly, the background factors related to the severity of anxiety and depressive symptoms were investigated. The factors that were significantly related to anxiety were insomnia and depression scores, while the factors associated with depression were life satisfaction, social involvement, and anxiety scores. Next, to identify the independent variables among these nutrients related to the severity of anxiety/depressive symptoms, a stepwise regression analysis was performed after eliminating multicollinearity. We found that the severity of both anxiety and depressive symptoms was only significantly associated with the intake of vitamin B6. Finally, a multivariate logistic regression analysis was performed to identify the independent relationships between the daily intake of vitamin B6 and moderate-to-severe anxiety/depressive symptoms. After adjusting for age, menopausal status (model 2), and background factors that were significantly related to the severity of anxiety/depressive symptoms (model 3), the intake of vitamin B6 was significantly associated with moderate-to-severe depression (model 2: adjusted odds ratio (AOR) per 10 μg/MJ in vitamin B6 intake = 0.91, 95% confidence interval (CI) = 0.85–0.97; model 3: AOR = 0.89, 95% CI = 0.80–0.99), while there was no significant relationship between the intake of vitamin B6 and the severity of anxiety after adjusting for the background variables (AOR = 0.97, 95% CI = 0.90–1.04; Table 4).

Table 1. The physical characteristics of all participants and comparison of these factors between women with mild, moderate, and severe anxiety and depressive symptoms.

Physical Characteristics	All Participants ($n = 289$)	Anxiety				Depression			
		Mild ($n = 156$)	Moderate ($n = 83$)	Severe ($n = 50$)	p-Value	Mild ($n = 178$)	Moderate ($n = 74$)	Severe ($n = 37$)	p-Value
Age (years)	52.0 (6.9)	54.3 (7.6)	52.6 (6.2)	52.3 (5.3)	0.279 [a]	54.7 (7.5)	52.2 (5.6)	50.2 (4.3)	0.002 [a]
Menopausal status (%)									
Pre/peri/postmenopausal	26.6/15.6/57.8	25.0/16.0/59.0	28.9/15.7/55.4	28.0/14.0/58.0	0.966 [b]	25.8/13.5/60.7	28.4/17.6/54.0	27.0/21.6/51.4	0.665 [b]
Body composition									
Height (cm)	157.0 (6.1)	156.6 (6.2)	157.5 (5.7)	155.8 (6.0)	0.206 [a]	156.4 (6.2)	157.6 (5.7)	156.6 (6.1)	0.313 [a]
Weight (kg)	52.5 (9.5)	53.6 (8.4)	54.8 (9.9)	54.6 (11.8)	0.822 [a]	53.6 (9.2)	55.0 (9.0)	54.8 (11.7)	0.377 [a]
Body mass index (kg/m^2)	21.3 (3.6)	21.9 (3.1)	22.1 (4.1)	22.4 (4.1)	0.834 [a]	21.9 (3.5)	22.2 (3.5)	22.3 (4.1)	0.801 [a]
Waist–hip ratio	0.9(0.1)	0.88 (0.06)	0.88 (0.06)	0.87 (0.07)	0.667 [a]	0.88 (0.06)	0.88 (0.06)	0.87 (0.07)	0.667 [a]
Fat mass (kg)	14.6 (6.9)	15.4 (6.0)	16.2 (7.6)	16.3 (8.2)	0.898 [a]	15.5 (6.7)	16.3 (6.7)	16.2 (8.4)	0.565 [a]
Lean mass (kg)	38.1 (3.7)	38.2 (3.6)	38.5 (3.5)	38.3 (4.4)	0.814 [a]	38.1 (3.7)	38.7 (3.4)	38.6 (4.3)	0.437 [a]
Muscle mass (kg)	35.9 (3.4)	36.0 (3.3)	36.3 (3.2)	36.1 (4.0)	0.811 [a]	35.9 (3.4)	36.5 (3.1)	36.4 (3.9)	0.436 [a]
Water mass (kg)	27.4 (3.3)	27.5 (3.1)	27.8 (3.2)	28.0 (4.1)	0.970 [a]	27.5 (3.3)	28.0 (3.3)	28.0 (3.7)	0.494 [a]
Basal metabolism (MJ/day)	4.59 (0.53)	4.61 (0.50)	4.68 (0.51)	4.65 (0.64)	0.748 [a]	4.60 (0.52)	4.70 (049)	4.69 (0.63)	0.314 [a]
Visceral fat level	5.0 (2.7)	5.3 (2.4)	5.4 (3.0)	5.4 (3.2)	0.962 [a]	5.3 (2.6)	5.4 (2.5)	5.2 (3.5)	0.492 [a]
Resting energy expenditure (MJ/day)	6.82 (1.85)	6.76 (1.88)	7.04 (1.91)	7.05 (1.69)	0.498 [a]	6.88 (1.98)	6.80 (1.55)	7.12 (1.84)	0.924 [a]
Body temprature (°C)	36.2 (0.6)	36.1 (0.7)	36.3 (0.5)	36.2 (0.5)	0.069 [a]	36.1 (0.7)	36.2 (0.5)	36.2 (0.5)	0.367 [a]
Physical fitness test									
Hand–grip strength (kg)	25.8 (4.9)	25.4 (4.8)	25.3 (4.6)	24.6 (5.3)	0.718 [a]	25.6 (4.5)	24.7 (5.1)	24.3 (5.8)	0.421 [a]
Ruler drop test (cm)	22.5 (4.3)	22.6 (4.3)	22.7 (4.1)	23.9 (4.7)	0.184 [a]	22.5 (4.1)	23.7 (4.1)	23.4 (5.4)	0.089 [a]
Sit-and-reach test (cm)	36.0 (10.1)	35.9 (10.4)	35.2 (10.4	36.0 (8.9)	0.763 [a]	36.0 (10.1)	35.6 (10.3)	34.7 (9.7)	0.658 [a]
Cardiovascular parameters									
Systolic blood pressure (mmHg)	125.5 (18.1)	124.4 (18.0)	127.7 (18.3)	128.8 (17.6)	0.172 [a]	126.8 (18.6)	125.5 (17.5)	124.2 (16.8)	0.666 [a]
Diastolic blood pressure (mmHg)	75.0 (12.4)	74.0 (11.6)	76.2 (12.6)	77.3 (14.2)	0.364 [a]	75.3 (11.5)	74.6 (13.3)	75.5 (14.7)	0.892 [a]
Heart rate (min^{-1})	77.0 (12.7)	78.6 (12.5)	78.2 (12.9)	78.6 (13.1)	0.908 [a]	78.3 (12.7)	80.3 (11.8)	76.0 (14.1)	0.043 [a]
Cardio–ankle vascular index	7.50 (0.78)	7.59 (0.76)	7.53 (0.79)	7.66 (0.82)	0.710 [a]	7.69 (0.79)	7.44 (0.73)	7.32 (0.73)	0.008 [a]
Ankle–brachial pressure index	1.11 (0.06)	1.12 (0.06)	1.11 (0.06)	1.09 (0.07)	0.251 [a]	1.11 (0.06)	1.11 (0.06)	1.10 (0.07)	0.663 [a]

Values are mean (standard deviation) or percentage. [a] Kruskal-Wallis test, [b] chi-squared test.

Table 2. The lifestyle and psychological characteristics of all the participants and comparison of these factors between women with mild, moderate, and severe anxiety and depressive symptoms.

Lifestyle and Psychological Characteristics	All Participants ($n = 289$)	Anxiety				Depression			
		Mild ($n = 156$)	Moderate ($n = 83$)	Severe ($n = 50$)	p-Value	Mild ($n = 178$)	Moderate ($n = 74$)	Severe ($n = 37$)	p-Value
Lifestyle factors									
Smoking (%)									
None/fewer than 20/20 or more cigarettes per day	93.1/3.1/3.8	91.0/3.9/5.1	96.4/2.4/1.2	94.0/2.0/4.0	0.563 [b]	92.7/2.2/5.1	97.3/2.7/0	86.5/8.1/5.4	0.487 [b]
Drinking (%)									
Never/sometimes/daily	66.8/24.6/8.7	68.6/23.7/7.7	61.5/28.9/9.6	70.0/20.0/10.0	0.739 [b]	64.6/27.5/7.9	70.3/21.6/8.1	70.3/16.2/13.5	0.934 [b]
Caffeine (%)									
Never/1–2 times/3 or more times per day	9.4/34.0/56.6	9.0/35.2/55.8	11.0/32.9/56.1	8.0/32.0/60.0	0.959 [b]	10.2/32.8/57.05.4/39.2/55.4	13.5/29.7/56.8		0.570 [b]
Exercise (%)									
Yes/No	49.0/51.0	58.7/41.3	42.2/57.8	30.0/70.0	<0.001 [b]	54.8/45.2	43.2/56.8	32.4/67.6	0.024 [b]
Menopausal Health–Related Quality of Life Questionnaire									
Somatic symptom score (0–18 points)	14.0 (4.0)	15.1 (3.9)	13.1 (3.6)	11.3 (3.7)	<0.001 [a]	15.0 (3.8)	12.3 (3.6)	11.6 (3.7)	<0.001 [a]
Vasomotor symptom score (0–6 points)	5.0 (2.2)	4.4 (2.0)	3.6 (2.3)	2.9 (2.0)	<0.001 [a]	4.3 (2.1)	3.3 (2.3)	3.4 (1.9)	<0.001 [a]
Insomnia symptom score (0–6 points)	5.0 (2.0)	4.7 (1.8)	3.6 (2.1)	2.8 (2.1)	<0.001 [a]	4.5 (1.9)	3.8 (2.1)	2.7 (2.2)	<0.001 [a]
Life satisfaction score (0–15 points)	8.0 (3.5)	8.8 (3.5)	6.9 (3.0)	5.6 (2.9)	<0.001 [a]	9.2 (2.0)	6.1 (2.7)	3.9 (2.5)	<0.001 [a]
Social involvement score (0–16 points)	9.0 (2.9)	9.7 (2.8)	9.1 (2.9)	8.5 (3.2)	0.060 [a]	10.2 (2.6)	8.4 (2.8)	7.1 (3.2)	<0.001 [a]
Hospital Anxiety and Depression Scale	12.0 (6.5)	8.8 (3.9)	16.6 (3.8)	22.4 (3.7)	<0.001 [a]	9.5 (4.2)	17.7 (3.1)	23.6 (3.8)	<0.001 [a]
Anxiety subscale score	7.0 (3.6)					5.5 (2.9)	8.9 (2.8)	10.9 (3.0)	<0.001 [a]
Depression subscale score	6.0 (3.6)	4.5 (2.8)	7.7 (3.4)	9.9 (2.9)	<0.001 [a]				
Psychotropic drug treatment (%)									
Yes/No	8.8/91.2	6.8/91.2	8.4/91.6	16.3/83.7	0.127 [b]	7.4/92.6	12.7/87.3	8.1/91.9	0.416 [b]

Values are mean (standard deviation) or percentage. [a] Kruskal-Wallis test, [b] chi-squared test.

Table 3. The nutritional intakes that differed significantly between the three anxiety/depression severity groups.

Nutrients	All Participants (n = 289)	Anxiety				Depression			
		Mild (n = 156)	Moderate (n = 83)	Severe (n = 50)	p-Value [a]	Mild (n = 178)	Moderate (n = 74)	Severe (n = 37)	p-Value [a]
Total energy (MJ)	6.99 (1.98)	6.83 (1.95)	7.43 (2.07)	6.76 (1.85)	0.083	6.85 (1.90)	7.42 (2.11)	6.78 (2.03)	0.167
Protein (%E)	16.3 (3.2)	16.6 (3.4)	15.9 (2.8)	16.2 (2.9)	0.236	16.8 (3.4)	15.5 (2.3)	15.9 (3.3)	0.048
Carbohydrate (%E)	52.7 (7.9)	52.3 (7.8)	53.2 (8.5)	52.9 (7.0)	0.724	51.7 (7.9)	54.7 (6.9)	53.0 (9.0)	0.027
Soluble dietary fiber (g/MJ)	0.50 (0.17)	0.51 (0.18)	0.46 (0.13)	0.51 (0.19)	0.155	0.52 (0.17)	0.46 (0.15)	0.45 (0.15)	0.014
Insoluble dietary fiber (g/MJ)	1.35 (0.43)	1.40 (0.46)	1.25 (0.33)	1.36 (0.47)	0.082	1.41 (0.45)	1.25 (0.38)	1.24 (0.36)	0.007
Dietary fiber (g/MJ)	1.90 (0.62)	1.97 (0.66)	1.77 (0.49)	1.93 (0.68)	0.102	2.00 (0.65)	1.77 (0.57)	1.74 (0.53)	0.006
Ash (g/MJ)	2.65 (0.53)	2.73 (0.58)	2.50 (0.40)	2.64 (0.50)	0.015	2.7 (0.6)	2.5 (0.4)	2.5 (0.6)	0.005
Potassium (mg/MJ)	397.1 (109.0)	413.4 (116.2)	366.2 (82.0)	397.4 (115.7)	0.015	417.2 (109.5)	367.1 (101.5)	360.3 (99.8)	<0.001
Calcium (mg/MJ)	85.5 (29.1)	103.3 (71.3)	79.9 (25.6)	84.1 (28.8)	0.017	89.8 (29.2)	78.6 (24.8)	78.7 (33.5)	0.002
Magnesium (mg/MJ)	37.7 (8.6)	41.3 (19.0)	35.7 (7.1)	38.0 (8.7)	0.029	39.2 (8.7)	35.3 (7.6)	36.7 (8.0)	0.001
Phosphorus (mg/MJ)	150.8 (32.4)	154.4 (34.6)	144.7 (29.2)	149.3 (29.3)	0.085	155.8 (33.9)	142.2 (26.0)	143.8 (32.4)	0.008
Iron (mg/MJ)	1.16 (0.30)	1.19 (0.31)	1.09 (0.23)	1.18 (0.31)	0.056	1.21 (0.30)	1.08 (0.26)	1.10 (0.29)	0.003
Zinc (mg/MJ)	1.12 (0.18)	1.13 (0.18)	1.10 (0.17)	1.10 (0.18)	0.271	1.15 (0.18)	1.08 (0.16)	1.08 (0.17)	0.007
β–Carotene (µg/MJ)	649.1 (469.1)	706.0 (544.3)	545.8 (304.5)	642.9 (415.6)	0.042	712.5 (526.5)	561.0 (341.1)	520.0 (329.9)	0.007
Retinol equivalent (µg/MJ)	109.2 (49.9)	114.3 (54.9)	97.0 (41.7)	113.3 (43.4)	0.030	114.4 (53.4)	99.0 (40.9)	104.0 (46.7)	0.110
Vitamin D (µg/MJ)	1.94 (1.31)	2.12 (1.41)	1.72 (1.00)	1.79 (1.40)	0.044	2.07 (1.40)	1.77 (0.95)	1.69 (1.43)	0.089
α-Tocopherol (mg/MJ)	1.09 (0.29)	1.13 (0.31)	1.03 (0.24)	1.10 (0.29)	0.056	1.13 (0.30)	1.04 (0.26)	0.99 (0.28)	0.006
Vitamin K (µg/MJ)	52.9 (27.1)	55.1 (29.4)	48.2 (20.5)	54.1 (28.7)	0.394	56.3 (28.5)	46.5 (22.5)	49.8 (26.7)	0.025
Vitamin B1 (mg/MJ)	0.11 (0.02)	0.12 (0.03)	0.11 (0.02)	0.11 (0.02)	0.068	0.12 (0.02)	0.11 (0.02)	0.10 (0.02)	<0.001
Vitamin B2 (mg/MJ)	0.20 (0.05)	0.20 (0.05)	0.19 (0.04)	0.20 (0.05)	0.088	0.20 (0.05)	0.18 (0.04)	0.19 (0.06)	0.005
Niacin (mgNE/MJ)	2.43 (0.63)	2.49 (0.63)	2.29 (0.60)	2.47 (0.66)	0.060	2.52 (0.62)	2.25 (0.53)	2.35 (0.78)	0.009
Vitamin B6 (µg/MJ)	185.3 (47.6)	192.0 (50.9)	169.6 (41.5)	182.9 (50.3)	0.005	193.7 (48.3)	167.3 (45.1)	171.3 (49.9)	<0.001
Folic acid (µg/MJ)	53.2 (20.1)	55.9 (21.8)	47.8 (14.9)	53.8 (20.8)	0.016	56.8 (21.0)	47.3 (17.3)	47.7 (16.9)	<0.001
Pantothenic acid (mg/MJ)	0.94 (0.19)	0.96 (0.19)	0.91 (0.18)	0.95 (0.20)	0.077	0.97 (0.19)	0.88 (0.17)	0.93 (0.22)	0.002
Vitamin C (mg/MJ)	18.9 (8.3)	20.0 (8.8)	16.8 (6.5)	19.0 (8.6)	0.020	20.3 (8.5)	17.2 (8.0)	15.4 (5.9)	<0.001

Values are mean (standard deviation). NE—niacin equivalent, [a] Kruskal–Wallis test.

Table 4. Associations between the daily intake of vitamin B6 (10 µg/MJ) and anxiety and depressive symptoms.

Model	Anxiety			Depressive		
	OR	95% CI	*p*-Value	OR	95% CI	*p*-Value
Model 1	0.93	0.88–0.98	0.004	0.89	0.84–0.94	<0.001
Model 2	0.93	0.88–0.99	0.016	0.91	0.85–0.97	0.002
Model 3	0.97	0.90–1.04	0.407	0.89	0.80–0.99	0.041

OR—odds ratio, CI—confidence interval. Model 1: unadjusted model. Model 2: multivariate logistic regression model, adjusted for age and menopausal status. Model 3: multivariate logistic regression model, adjusted for age, menopausal status, and the background factors related to the severity of the anxiety/depressive symptoms (insomnia and depression scores/life satisfaction, social involvement, and anxiety scores).

4. Discussion

In our cross-sectional analysis, the severity of depressive symptoms was significantly inversely associated with the dietary intake of vitamin B6 in Japanese middle-aged and elderly women. Vitamin B6, which comprises six chemical compounds (pyridoxine, pyridoxamine, pyridoxal, and their phosphorylated derivatives pyridoxine 5′-phosphate, pyridoxamine 5′-phosphate, and pyridoxal 5′-phosphate (PLP)), is richly contained in red pepper, garlic, nuts, fish, and meats. PLP, the most active form, serves as an enzymatic cofactor in more than 140 different biochemical reactions, such as those involving amino acids, neurotransmitters, heme biosynthesis, fatty acid metabolism, and glycogen breakdown [27,28].

It is well known that neurotransmitters, such as serotonin, dopamine, norepinephrine, γ-aminobutyric acid, and glutamate, play a critical role in the development of psychiatric disorders, and their receptors could be potential therapeutic targets for the treatment of psychoneurological symptoms. Abundant reports have shown that the dysregulation of monoamine systems contributes to anxiety and depressive disorders [29,30]. Serotonin is synthesized from tryptophan by PLP-dependent dopa decarboxylase, and dopamine and norepinephrine production are also required for the catalysis of PLP-dependent dopa decarboxylase. Decreased vitamin B6 (PLP) could be associated with monoamine depletion and impaired neurotransmission. Furthermore, the kynurenine pathway involving two PLP-dependent enzymes, which is a major tryptophan metabolic pathway, is associated with depression [31]. The disturbance of the balance between the neuroprotection and neurotoxicity of kynurenine pathway metabolites—i.e., kynurenines, such as kynurenic acid and quinolinic acid—plays a key role in the development of depression [32]. It is supposed that PLP-dependent kynureninase is more sensitive to PLP deficiency than is the PLP-dependent kynurenine aminotransferase; thus, PLP deficiency reduces kynureninase activity first [33,34]. Therefore, kynurenic acid, 3-hydroxykynurenine, and xanthurenic acid could increase, although, the changes in the plasma and urinary levels of kynurenines via vitamin B6 depletion are inconsistent [33–35]. Kynurenic acid inhibits N-methyl-D-aspartate receptors and alfa-acetylcholine receptors in the central nervous system [36,37], leading to a decline in the extracellular levels of acetylcholine, glutamate, and dopamine. Additionally, 3-hydroxykynurenine, which is a redox-active metabolite, is neurotoxic through generating reactive oxygen species and eventually induces apoptosis [38], while xanthurenic acid acts as a metabolic glutamate 2/3 receptor agonist, which could improve positive and negative symptoms in schizophrenia [39]. Vitamin B6 depletion might cause a dysregulated neurotransmitter system and neural dysfunction through imbalances of tryptophan metabolites via kynurenine.

There are several reports on the association between vitamin B6 and depression. Hvas and colleagues showed that low plasma levels of PLP were related to depressive symptoms in 140 study participants [40]. A seven-year longitudinal study of 3503 adults aged ≥65 years demonstrated that a higher total intake of vitamin B6 (dietary and supplementary intake) was related to a lower likelihood of depression [41]. Moreover, a higher dietary intake of vitamin B6 was associated with a lower incidence of depression among women in a three-year longitudinal study of 1793 adults aged ≥68 years [42]. Additionally, a few systematic reviews of the effects on mood of vitamin B6 alone, or a combinative intervention of vitamins and minerals, such as folate, vitamin B12, vitamin C, vitamin D, magnesium, calcium, and iron, on mood supported the idea that supplementation with B6 vitamins could relieve mood symptoms. For example, Williams and colleagues reported the beneficial effects of vitamin B6 supplementation on depression among premenopausal women [43]. Young and colleagues also revealed that the supplementation of B vitamins might alleviate mood symptoms in healthy adults and adults at risk for mental disorders [44].

In contrast, several randomized controlled trials failed to find significant effects due to a combinative intervention of B vitamins, including vitamin B6, on mood symptoms [45–47]. In the current study, the mean daily vitamin B6 intake was smaller than the recommended dietary allowance only in the severe depressive group [48], which might affect our results. Further studies should be conducted to determine the exact effects of vitamin B6 as an independent treatment.

The major limitations of our study were the relatively small sample size and uncertain causal relationship owing to its cross-sectional nature. It may not be appropriate to generalize our findings to a wider population. We did not investigate the serum levels of vitamin B6, although we estimated the daily intake of vitamin B6 using the BDHQ. Therefore, it was uncertain whether the severity of depression was related to serum vitamin B6 levels. Furthermore, the BDHQ, which is a method based on food recall to determine the frequency of food eaten, provided information only for the 61 listed foods and beverages. In addition, a potential contributor to mood, namely, the use of dietary supplements, was not assessed. The use of dietary supplements, such as vitamins (B1, B2, B6, C, and E) and minerals (calcium and iron), has been estimated at only 7.7% in Japanese women [49]. Nevertheless, the dietary intake in this study did not represent the total nutrient intake.

Nonetheless, our study has several strengths and novel features. As many as 43 nutrients and various background factors, including physical and psychological health status, life satisfaction, and social involvement, were analyzed simultaneously. Therefore, we found that the intake of vitamin B6 was independently associated with the severity of depressive symptoms. To the best of our knowledge, this is the first report on the relationship between the intake of vitamin B6 and depressive symptoms as a result of an analysis of various nutrients.

In conclusion, moderate-to-severe depressive symptoms were associated with a lower dietary intake of vitamin B6 in Japanese middle-aged and elderly women. A higher intake of vitamin B6 could help relieve depressive symptoms in this population.

Supplementary Materials: The following are available online at http://www.mdpi.com/2072-6643/12/11/3437/s1, Table S1: The major 43 nutrients assessed by the brief self-administered diet history questionnaire, Table S2: The Menopausal Health-Related Quality of Life Questionnaire, Table S3: The differences in daily intake of nutrients between the severity groups.

Author Contributions: T.O. and M.T. were responsible for the project development, data collection, and data analysis. R.S. contributed to data analysis and K.K. was responsible for data collection. A.H. and N.M. participated in the project development and supervision. All authors have read and agreed to the published version of the manuscript.

Funding: This research was funded by the Kikkoman Corporation, grant number x2136.

Conflicts of Interest: M.T. received an unrestricted research grant from the Kikkoman Corporation. The other authors disclose no conflicts of interest. The authors alone are responsible for the content and writing of the paper. The funders had no role in the design of the study; in the collection, analyses, or interpretation of data; in the writing of the manuscript, or in the decision to publish the results.

References

1. World Health Organization: Depression and Other Common Mental Disorders, Global Health Estimates. Available online: https://apps.who.int/iris/bitstream/handle/10665/254610/WHO-MSD-MER-2017.2-eng.pdf#search=%27WHO+depression%27 (accessed on 1 October 2020).
2. Noble, R.E. Depression in women. *Metabolism* **2005**, *54*, 49–52. [CrossRef] [PubMed]

3. Ryan, J.; Burger, H.G.; Szoeke, C.; Lehert, P.; Ancelin, M.L.; Henderson, V.W.; Dennerstein, L. A prospective study of the association between endogenous hormones and depressive symptoms in postmenopausal women. *Menopause* **2009**, *16*, 509–517. [CrossRef] [PubMed]

4. Vivian-Taylor, J.; Hickey, M. Menopause and depression: Is there a link? *Maturitas* **2014**, *79*, 142–146. [CrossRef] [PubMed]

5. Freeman, E.W.; Sammel, M.D.; Lin, H.; Nelson, D.B. Associations of hormones and menopausal status with depressed mood in women with no history of depression. *Arch. Gen. Psychiatry* **2006**, *63*, 375–382. [CrossRef] [PubMed]

6. Cohen, L.S.; Soares, C.N.; Vitonis, A.F.; Otto, M.W.; Harlow, B.L. Risk for new onset of depression during the menopausal transition: The Harvard study of moods and cycles. *Arch. Gen. Psychiatry* **2006**, *63*, 385–390. [CrossRef] [PubMed]

7. Georgakis, M.K.; Thomopoulos, T.P.; Diamantaras, A.A.; Kalogirou, E.I.; Skalkidou, A.; Daskalopoulou, S.S.; Petridou, E.T. Association of age at menopause and duration of reproductive period with depression after menopause: A systematic review and meta-analysis. *JAMA Psychiatry* **2016**, *73*, 139–149. [CrossRef]

8. Deecher, D.; Andree, T.H.; Sloan, D.; Schechter, L.E. From menarche to menopause: Exploring the underlying biology of depression in women experiencing hormonal changes. *Psychoneuroendocrinology* **2008**, *33*, 3–17. [CrossRef]

9. Bromberger, J.T.; Schott, L.L.; Kravitz, H.M.; Sowers, M.; Avis, N.E.; Gold, E.B.; Randolph, J.F., Jr.; Matthews, K.A. Longitudinal change in reproductive hormones and depressive symptoms across the menopausal transition: Results from the Study of Women's Health Across the Nation (SWAN). *Arch. Gen. Psychiatry* **2010**, *67*, 598–607. [CrossRef]

10. Zweifel, J.E.; O'Brien, W.H. A meta-analysis of the effect of hormone replacement therapy upon depressed mood. *Psychoneuroendocrinology* **1997**, *22*, 189–212. [CrossRef]

11. Rudolph, I.; Palombo-Kinne, E.; Kirsch, B.; Mellinger, U.; Breitbarth, H.; Gräser, T. Influence of a continuous combined HRT (2 mg estradiol valerate and 2 mg dienogest) on postmenopausal depression. *Climacteric* **2004**, *7*, 301–311. [CrossRef]

12. Gleason, C.E.; Dowling, N.M.; Wharton, W.; Manson, J.E.; Miller, V.M.; Atwood, C.S.; Brinton, E.A.; Cedars, M.I.; Lobo, R.A.; Merriam, G.R.; et al. Effects of hormone therapy on cognition and mood in recently postmenopausal women: Findings from the randomized, controlled KEEPS-Cognitive and Affective Study. *PLoS Med.* **2015**, *12*, e1001833. [CrossRef] [PubMed]

13. Morrison, M.F.; Kallan, M.J.; Ten Have, T.; Katz, I.; Tweedy, K.; Battistini, M. Lack of efficacy of estradiol for depression in postmenopausal women: A randomized, controlled trial. *Biol. Psychiatry* **2004**, *55*, 406–412. [CrossRef] [PubMed]

14. Benton, D.; Donohoe, R.T. The effects of nutrients on mood. *Public Health Nutr.* **1999**, *2*, 403–409. [CrossRef] [PubMed]

15. LaChance, L.R.; Ramsey, D. Antidepressant foods: An evidence-based nutrient profiling system for depression. *World J. Psychiatry* **2018**, *8*, 97–104. [CrossRef] [PubMed]

16. Zigmond, A.S.; Snaith, R.P. The hospital anxiety and depression scale. *Acta Psychiatr. Scand.* **1983**, *67*, 361–370. [CrossRef] [PubMed]

17. National Health and Nutrition Survey, Intake by Food Groups. Available online: https://www.nibiohn.go.jp/eiken/kenkounippon21/en/eiyouchousa/kekka_syokuhin_chousa_koumoku.html (accessed on 20 July 2020).

18. Kobayashi, S.; Murakami, K.; Sasaki, S.; Okubo, H.; Hirota, N.; Notsu, A.; Fukui, M.; Date, C. Comparison of relative validity of food group intakes estimated by comprehensive and brief-type self-administered diet history questionnaires against 16 d dietary records in Japanese adults. *Public Health Nutr.* **2011**, *14*, 1200–1211. [CrossRef]

19. Kobayashi, S.; Honda, S.; Murakami, K.; Sasaki, S.; Okubo, H.; Hirota, N.; Notsu, A.; Fukui, M.; Date, C. Both comprehensive and brief self-administered diet history questionnaires satisfactorily rank nutrient intakes in Japanese adults. *J. Epidemiol.* **2012**, *22*, 151–159. [CrossRef]

20. van Keep, P.A.; Kellerhals, J.M. The impact of socio-cultural factors on symptom formation. Some results of a study on ageing women in Switzerland. *Psychother Psychosom.* **1974**, *23*, 251–263. [CrossRef]

21. Hunter, M.; Battersby, R.; Whitehead, M. Relationships between psychological symptoms, somatic complaints and menopausal status. *Maturitas* **1986**, *8*, 217–228. [CrossRef]

22. Hanson, B.S.; Isacsson, S.O.; Janzon, L.; Lindell, S.E. Social network and social support influence mortality in elderly men. The prospective population study of "Men born in 1914," Malmö, Sweden. *Am. J. Epidemiol.* **1989**, *130*, 100–111. [CrossRef]

23. Terauchi, M.; Obayashi, S.; Akiyoshi, M.; Kato, K.; Matsushima, E.; Kubota, T. Insomnia in Japanese peri- and postmenopausal women. *Climacteric* **2010**, *13*, 479–486. [CrossRef] [PubMed]

24. Terauchi, M.; Hiramitsu, S.; Akiyoshi, M.; Owa, Y.; Kato, K.; Obayashi, S.; Matsushima, E.; Kubota, T. Associations between anxiety, depression and insomnia in peri- and post-menopausal women. *Maturitas* **2012**, *72*, 61–65. [CrossRef] [PubMed]

25. Terauchi, M.; Hirose, A.; Akiyoshi, M.; Owa, Y.; Kato, K.; Kubota, T. Prevalence and predictors of storage lower urinary tract symptoms in perimenopausal and postmenopausal women attending a menopause clinic. *Menopause* **2015**, *22*, 1084–1090. [CrossRef] [PubMed]

26. Hirose, A.; Terauchi, M.; Akiyoshi, M.; Owa, Y.; Kato, K.; Kubota, T. Subjective insomnia is associated with low sleep efficiency and fatigue in middle-aged women. *Climacteric* **2016**, *19*, 369–374. [CrossRef] [PubMed]

27. Mooney, S.; Leuendorf, J.E.; Hendrickson, C.; Hellmann, H. Vitamin B6: A long known compound of surprising complexity. *Molecules* **2009**, *14*, 329–351. [CrossRef] [PubMed]

28. Parra, M.; Stahl, S.; Hellmann, H. Vitamin B6 and its role in cell metabolism and physiology. *Cells* **2018**, *7*, 84. [CrossRef] [PubMed]

29. Ressler, K.J.; Nemeroff, C.B. Role of serotonergic and noradrenergic systems in the pathophysiology of depression and anxiety disorders. *Depress. Anxiety* **2000**, *12*, 2–19. [CrossRef]

30. Grace, A.A. Dysregulation of the dopamine system in the pathophysiology of schizophrenia and depression. *Nat. Rev. Neurosci.* **2016**, *17*, 524–532. [CrossRef]

31. Curzon, G.; Bridges, P.K. Tryptophan metabolism in depression. *J. Neurol. Neurosurg. Psychiatry* **1970**, *33*, 698–704. [CrossRef]

32. Myint, A.M.; Kim, Y.K. Cytokine-serotonin interaction through IDO: A neurodegeneration hypothesis of depression. *Med. Hypotheses* **2003**, *61*, 519–525. [CrossRef]

33. Ogasawara, N.; Hagino, Y.; Kotake, Y. Kynurenine-transaminase, kynureninase and the increase of xanthurenic acid excretion. *J. Biochem.* **1962**, *52*, 162–166. [CrossRef] [PubMed]

34. Ueland, P.M.; McCann, A.; Midttun, Ø.; Ulvik, A. Inflammation, vitamin B6 and related pathways. *Mol. Aspects Med.* **2017**, *53*, 10–27. [CrossRef] [PubMed]

35. Shibata, K.; Mushiage, M.; Kondo, T.; Hayakawa, T.; Tsuge, H. Effects of vitamin B6 deficiency on the conversion ratio of tryptophan to niacin. *Biosci. Biotechnol. Biochem.* **1995**, *59*, 2060–2063. [CrossRef] [PubMed]

36. Stone, T.W. Neuropharmacology of quinolinic and kynurenic acids. *Pharmacol. Rev.* **1993**, *45*, 309–379.

37. Hilmas, C.; Pereira, E.F.; Alkondon, M.; Rassoulpour, A.; Schwarcz, R.; Albuquerque, E.X. The brain metabolite kynurenic acid inhibits alpha7 nicotinic receptor activity and increases non-alpha7 nicotinic receptor expression: Physiopathological implications. *J. Neurosci.* **2001**, *21*, 7463–7473. [CrossRef]

38. Okuda, S.; Nishiyama, N.; Saito, H.; Katsuki, H. 3-Hydroxykynurenine, an endogenous oxidative stress generator, causes neuronal cell death with apoptotic features and region selectivity. *J. Neurochem.* **1998**, *70*, 299–307. [CrossRef]

39. Patil, S.T.; Zhang, L.; Martenyi, F.; Lowe, S.L.; Jackson, K.A.; Andreev, B.V.; Avedisova, A.S.; Bardenstein, L.M.; Gurovich, I.Y.; Morozova, M.A.; et al. Activation of mGlu2/3 receptors as a new approach to treat schizophrenia: A randomized Phase 2 clinical trial. *Nat. Med.* **2007**, *13*, 1102–1107. [CrossRef]

40. Hvas, A.M.; Juul, S.; Bech, P.; Nexø, E. Vitamin B6 level is associated with symptoms of depression. *Psychother Psychosom.* **2004**, *73*, 340–343. [CrossRef]

41. Skarupski, K.A.; Tangney, C.; Li, H.; Ouyang, B.; Evans, D.A.; Morris, M.C. Longitudinal association of vitamin B-6, folate, and vitamin B-12 with depressive symptoms among older adults over time. *Am. J. Clin. Nutr.* **2010**, *92*, 330–335. [CrossRef]

42. Gougeon, L.; Payette, H.; Morais, J.A.; Gaudreau, P.; Shatenstein, B.; Gray-Donald, K. Intakes of folate, vitamin B6 and B12 and risk of depression in community-dwelling older adults: The Quebec Longitudinal Study on Nutrition and Aging. *Eur. J. Clin. Nutr.* **2016**, *70*, 380–385. [CrossRef]

43. Williams, A.L.; Cotter, A.; Sabina, A. The role for vitamin B-6 as treatment for depression: A systematic review. *Fam. Pract.* **2005**, *22*, 532–537. [CrossRef] [PubMed]

44. Young, L.M.; Pipingas, A.; White, D.J.; Gauci, S.; Scholey, A. A systematic review and meta-analysis of B vitamin supplementation on depressive symptoms, anxiety, and stress: Effects on healthy and 'at-risk' individuals. *Nutrients* **2019**, *11*, 2232. [CrossRef] [PubMed]

45. Bryan, J.; Calvaresi, E.; Hughes, D. Short-term folate, vitamin B-12 or vitamin B-6 supplementation slightly affects memory performance but not mood in women of various ages. *J. Nutr.* **2002**, *132*, 1345–1356. [CrossRef] [PubMed]

46. Okereke, O.I.; Cook, N.R.; Albert, C.M.; van Denburgh, M.; Manson, G.E. Effect of long-term supplementation with folic acid and B vitamins on risk of depression in older women. *Br. J. Psychiatry* **2015**, *206*, 324–331. [CrossRef] [PubMed]

47. De Koning, E.J.; van der Zwaluw, N.L.; van Wijngaarden, J.P.; Sohl, E.; Brouwer-Brolsma, E.M.; van Marwijk, H.W.; Enneman, A.W.; Swart, K.M.; van Dijk, S.C.; Ham, A.C.; et al. Effects of two-year vitamin B12 and folic acid supplementation on depressive symptoms and quality of life in older adults with elevated homocysteine concentrations: Additional results from the B-PROOF Study, an RCT. *Nutrients* **2016**, *8*, 748. [CrossRef]

48. Ministry of Health, Labor and Welfare in Japan. Overview of Dietary Reference Intakes for Japanese (2015). Available online: https://www.mhlw.go.jp/file/06-Seisakujouhou-10900000-Kenkoukyoku/Overview.pdf (accessed on 21 July 2020).

49. Tsubota-Utsugi, M.; Nakade, M.; Imai, E.; Tsuboyama-Kasaoka, N.; Nozue, M.; Umegaki, K.; Yoshizawa, T.; Okuda, N.; Nishi, N.; Takimoto, H. Distribution of vitamin E intake among Japanese dietary supplement and fortified food users: A secondary analysis from the National Health and Nutrition Survey, 2003–2009. *J. Nutr. Sci. Vitaminol. (Tokyo)* **2013**, *59*, 576–583. [CrossRef] [PubMed]

Publisher's Note: MDPI stays neutral with regard to jurisdictional claims in published maps and institutional affiliations.

© 2020 by the authors. Licensee MDPI, Basel, Switzerland. This article is an open access article distributed under the terms and conditions of the Creative Commons Attribution (CC BY) license (http://creativecommons.org/licenses/by/4.0/).

Article

Effects of Chlorogenic Acids on Menopausal Symptoms in Healthy Women: A Randomized, Placebo-Controlled, Double-Blind, Parallel-Group Trial

Yuka Enokuchi [1,*], Atsushi Suzuki [2], Tohru Yamaguchi [1], Ryuji Ochiai [3], Masakazu Terauchi [4] and Kiyoshi Kataoka [5]

1   Health & Wellness Products Research Laboratories, Kao Corporation, Tokyo 131-8501, Japan; yamaguchi.tohru@kao.com
2   Department of Research & Development, Kao Corporation, Tokyo 131-8501, Japan; suzuki.atsushi2@kao.com
3   Biological Science Research Laboratories, Kao Corporation, Tokyo 131-8501, Japan; ochiai.ryuuji@kao.com
4   Department of Woman's Health, Tokyo Medical and Dental University, Tokyo 113-8510, Japan; teragyne@tmd.ac.jp
5   Personal Health Care Products Research Laboratories, Kao Corporation, Tokyo 131-8501, Japan; kataoka.kiyoshi@kao.com
*   Correspondence: enokuchi.yuka@kao.com; Tel.: +81-3-5630-9877

Received: 30 October 2020; Accepted: 4 December 2020; Published: 7 December 2020

Abstract: A reduction in estrogen levels in the perimenopausal and postmenopausal periods causes various symptoms in women, such as hot flushes, sweats, depression, anxiety, and insomnia. Chlorogenic acids (CGAs), which are phenolic compounds widely present in plants such as coffee beans, have various physiological functions. However, the effects of CGAs on menopausal symptoms are unknown. To examine the effects of CGAs on menopausal symptoms, especially hot flushes, a randomized, placebo-controlled, double-blind, parallel-group trial was conducted in healthy women. Eighty-two subjects were randomized and assigned to receive CGAs (270 mg) tablets or the placebo for 4 weeks. After 4 weeks of intake, the number of hot flushes, the severity of hot flushes during sleep, and the severity of daytime sweats decreased significantly in the CGA group compared to the placebo group. The modified Kupperman index for menopausal symptoms decreased significantly after 2 weeks in the CGA group compared to the placebo group. Adverse effects caused by CGAs were not observed. The results show that continuous intake of CGAs resulted in improvements in menopausal symptoms, especially hot flushes, in healthy women.

Keywords: chlorogenic acid; hot flushes; menopausal symptoms

---

1. Introduction

Peri- and postmenopausal women are afflicted by various physical and psychological disorders that are collectively referred to as menopausal symptoms, including hot flushes, sweats, insomnia, and anxiety. Hot flushes and sweats are considered vasomotor symptoms because of their vascular reactivity with vasodilation and vasoconstriction, and usually start from the face—mainly the upper body—and may spread to the head, chest, and even the whole body. Vasomotor symptoms interfere with daily activities and sleep, and can cause fatigue, loss of concentration, and symptoms of depression, affecting the quality of life (QOL) of menopausal women. Several factors have been reported to exacerbate vasomotor symptoms, for example, surgical menopause, high body mass index (BMI), and smoking.

The mechanism underlying the development of vasomotor symptoms is not fully understood. One hypothesis is that menopausal women with hot flushes have a narrower thermoneutral zone [1]. Kisspeptin/neurokinin B/dynorphin (KNDy) neurons [2], calcitonin gene-related peptide [3], vascular endothelial function [4], and autonomic nervous system function [5,6] have also been reported to be involved in the development of vasomotor symptoms. The domino theory [7] suggests that hot flushes result in reduced sleep duration, leading to the exacerbation of psychological symptoms due to a lack of sleep. Conversely, improving sleep conditions may lead to a decrease in hot flushes [8].

Various methods have been reported to relieve menopausal symptoms, and hormone replacement therapy (HRT) is accepted as the most effective treatment for hot flushes. Despite concerns about the safety of HRT after the publication of the Women's Health Initiative (WHI) study [9], HRT has many functions in improving women's QOL, and is considered the most effective treatment [10]. Some types of *kampo* (Japanese traditional herbal medicine) have been reported to improve menopausal symptoms [11,12], and herbs and dietary nutrients such as black cohosh [13], isoflavone [14,15], equol (a metabolite of isoflavones) [16], and proanthocyanidin [17] are also known to relieve menopausal symptoms.

Previous research has suggested that racial differences exist in the severity of hot flushes, with Asian women experiencing fewer hot flushes than Caucasian women [18]. In the equol study [16], subjects were Japanese women who experienced hot flushes approximately three times a day on average; however, dietary nutrients that reduce the relatively infrequent hot flushes may be more effective for Japanese women during menopause.

Chlorogenic acids (CGAs) are phenolic compounds widely present in plants such as coffee beans. CGAs have been documented to have antitumor [19] and antioxidant [20] effects in addition to effects in improving vascular endothelial function and hypertension [21] and reducing body fat [22]. Recent studies have reported that the continuous intake of CGAs improves autonomic nervous system activity [23] and sleep quality [24]; however, the effects on women's menopausal symptoms are unknown. In the present study, the effects of the continuous intake of CGAs on menopausal symptoms, especially hot flushes, were examined in healthy menopausal women.

2. Materials and Methods

2.1. Study Design

A randomized, placebo-controlled, double-blind, parallel-group trial was conducted to assess the effects of CGA intake on menopausal symptoms, especially hot flushes. The total test period was 6 weeks, including a pre-observation period of 1 week (week 0), a test tablet intake period of 4 weeks (weeks 1–4), and a post-observation period of 1 week (week 5). Subjects took either 6 CGA tablets containing green coffee bean extract or 6 placebo tablets with water before sleep.

The study protocol was prepared prior to commencing the study. The study was conducted after review and approval by the Human Research Ethics Committee of Kao Corporation (approval no. T150-180720, approved 23 August 2018) and registration with the University Hospital Medical Information Network (UMIN) Clinical Trials Registry (ID: UMIN000034056). The study was conducted from September 2018 to April 2019 in accordance with the Declaration of Helsinki and directed entirely by Research and Development Inc. (Tokyo, Japan). The primary endpoint was menopausal symptoms (the number of hot flushes) at four weeks after the completion of intake, and the secondary endpoints were the number of sweats, severities of hot flushes and sweats, general menopausal symptoms, health-related QOL (HRQOL), and anxiety.

2.2. Subjects

Healthy female volunteers aged 40–59 years with menopausal symptoms were recruited from around the Tokyo metropolitan area. The inclusion criteria for the subjects were as follows: healthy women 40–59 years of age, moderate or severe rating according to the modified Kupperman menopausal index (mKMI) severity grading (including a moderate or a lower score in some cases), and subjective symptoms of hot flushes. The exclusion criteria were as follows: those affected by disease; taking regular medication; receiving regular outpatient treatment at a medical facility or visited a hospital within one month prior to the pre-examination; taking HRT; taking medication that may interfere with hormones; regularly taking supplements that affect menopausal symptoms; regularly taking functional foods and supplements that have been shown to be effective in improving sleep; pregnant, breastfeeding, or willing to become pregnant during the study period; smokers; unable to follow the dietary restrictions during the study period; known to have a past or current history of allergies; leading irregular lifestyles; or deemed unfit for enrollment by the physician in charge or by the principal investigator due to other causes. During the study, subjects limited their coffee intake to one cup per day. Written informed consent was obtained from all subjects before the commencement of the study.

Figure 1 shows the flow from subject enrollment to analysis. Of the 90 recruited subjects, 82 participated in the study. These subjects were randomly assigned to two groups, the CGA group and the placebo group, using stratified randomization to ensure that the subjects were equally distributed (n = 41 subjects per group) according to the number of hot flushes, severity of menopausal symptoms, age, and menstrual status as identified in the pre-survey.

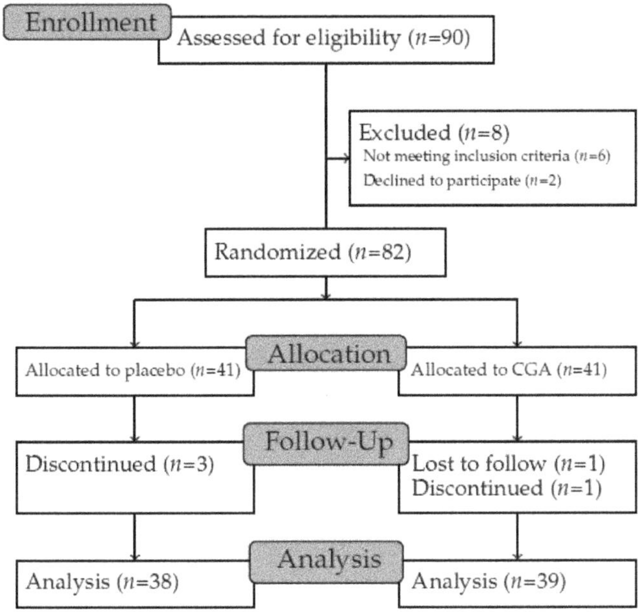

Figure 1. Flow diagram of the study from subject enrollment to analysis. CGA, chlorogenic acid.

Of the 82 subjects, three in the placebo group withdrew from the study (one subject was prescribed an antihypertensive drug during the pre-observation period, one subject developed symptoms of insomnia during the study period, and one subject was hospitalized due to an injury during the study period). In the CGA group, two subjects were excluded from the study (one subject discontinued the intake of the test tablets due to a cold during the study period, and one subject for personal reasons). Therefore, 77 subjects who completed study (38 subjects in the placebo group and 39 subjects in the CGA group) were included in the analysis. The mean intake rate of the test tablets among those who completed the study was 98.6% in the CGA group and 99.6% in the placebo group. For the 385 data points (corresponding to 5 weeks multiplied by 77 subjects), there were three missing measurements in the hot flush dataset and six missing measurements in the sweat dataset.

The background characteristics of the subjects who completed the study in each group are shown in Table 1. There were no significant differences between the two groups in terms of any of the items. The causal relationships of adverse events identified during and after the study was completed were evaluated by a physician. Of the four adverse events identified during the study period, the causal relationship with the study intervention of one subject in the placebo group, who developed insomnia after the start of ingestion of the study tablets, was deemed "undeniable"; however, there was no conclusion that there was a causal relationship, including the events identified after the completion of the study.

Table 1. Baseline characteristics by treatment group (per protocol set).

	Placebo	CGA	Test Statistic
	$n = 38$	$n = 39$	
Age (years)	48.2, **51.0**, 55.0 (51.4 ± 4.3)	48.0, **51.0**, 54.5 (51.3 ± 4.6)	$F_{1,75} = 0$, $p = 0.992$ [1]
Height (cm)	155.0, **159.5**, 162.0 (158.4 ± 5.0)	155.5, **158.0**, 162.0 (158.5 ± 4.7)	$F_{1,75} = 0.02$, $p = 0.899$ [1]
Body weight (kg)	49.2, **54.5**, 60.0 (55.0 ± 7.2)	51.0, **55.0**, 59.0 (54.9 ± 6.9)	$F_{1,75} = 0.01$, $p = 0.932$ [1]
BMI (kg/m^2)	19.6, **21.7**, 23.7 (22.0 ± 3.1)	19.9, **21.3**, 23.2 (21.9 ± 2.8)	$F_{1,75} = 0.03$, $p = 0.872$ [1]
mKMI total score	17.2, **22.0**, 27.8 (23.1 ± 8.0)	16.0, **22.0**, 27.5 (22.5 ± 8.0)	$F_{1,75} = 0.12$, $p = 0.731$ [1]
Menopause Status			$\chi^2_3 = 1.25$, $p = 0.742$ [2]
Premenopausal	16% (6)	15% (6)	
Perimenopausal	32% (12)	36% (14)	
Postmenopausal	53% (20)	46% (18)	
Missing	0% (0)	3% (1)	

The three numbers left **center** right represent the lower quartile left, the median **center**, and the upper quartile right for continuous variables. x ± s represents $\overline{X}$ ± 1 SD. Numbers in parentheses after percentages indicate frequencies. Tests used: [1] Wilcoxon rank-sum test; [2] Pearson's χ^2 test. CGA, chlorogenic acid; BMI, body mass index; mKMI, modified Kupperman menopausal index.

2.3. Test Tablets

CGAs were extracted from green coffee beans with hot water, and caffeine was removed from the extract using activated carbon. The extract was spray-dried to obtain a powder. The compositions of the CGAs were determined using high-performance liquid chromatography and were as follows: caffeoylquinic acids (CQAs) 5-caffeoylquinic acid, 3-caffeoylquinic acid, and 4-caffeoylquinic acid; feruloylquinic acids (FQAs) 3-feruloylquinic acid, 4-feruloylquinic acid, and 5-feruloylquinic acid; dicaffeoylquinic acids (di-CQAs) 3,4-dicaffeoylquinic acid, 3,5-dicaffeoylquinic acid, and 4,5-dicaffeoylquinic acid. The CGA tablets (300 mg/tablet, 6 tablets/day) were industrially manufactured to be homogeneous with excipients and flavoring so that six CGA tablets contained 270 mg of the main CGA components (CQAs and FQAs). The placebo tablets did not contain CGAs and were replaced with excipients. According to an analysis by a third-party analysis agency for manufactured tablets, the six CGA tablets contained 272.5 mg of the main components of CGAs (CQAs and FQAs), 1.35 g of carbohydrates (including CGAs), 338 mg of protein, and 34.2 mg of fat. The six placebo tablets contained 1.72 g of carbohydrates, 1.8 mg of protein, and 37.8 mg of fat. The subjects could not identify which tablets they were taking based on the appearance and taste of the tablets.

2.4. Assessment of Hot Flushes and Sweats

The subjects recorded the number of hot flushes and sweats, wake-up time, and bedtime over a total period of 6 weeks (weeks 0–5), and they also recorded whether test tablets were taken during weeks 1–4 (4 weeks) in their daily diary. The severities of hot flushes and sweats during the daytime or sleep was assessed every two weeks using a visual analog scale (VAS) (last day of weeks 0, 2, and 4). The VAS was represented as a 100 mm-long straight line with the left end representing "no symptoms at all" and the right end representing "extremely severe symptoms". Subjects marked the scale according to the severity of items, and the position was measured as the distance (mm) from the left edge.

2.5. Assessment of Menopausal Symptoms, HRQOL, and Anxiety

The mKMI [25] is a questionnaire that is a modification of the Kupperman menopausal index (KMI) [26] for the assessment of Japanese women. The mKMI includes 17 questions in which questions related to symptoms characteristic of Japanese menopausal women are added to the 11 questions of the original KMI. The 17 questions were classified into the following 11 symptoms according to the KMI: vasomotor, paresthesia, insomnia, nervousness, melancholia, vertigo, weakness/fatigue, arthralgia and myalgia, headaches, palpitation, and formication. For each of the 17 questions, subjects selected the level of the condition according to 4 levels (strong, medium, weak, and none) experienced in the previous 2–3 days. The highest severity for each question in the symptom group was taken as the severity of the symptom group, and the score was calculated using the KMI weighting method. The highest possible total score for the mKMI was 51, which is the same as for the KMI, and the severity grading was assessed according to the following scores: 0–12 is I, 13–22 is II, 23–33 is III, 34–43 is IV, and 44–51 is V. I and II are considered as mild, III as moderate, and IV and V as severe. The mKMI was used to evaluate the effect of intervention, and determining the severity was not its primary purpose. We used the mKMI to confirm the effects of the intervention.

The Short Form-8 (SF-8) is a scale that can measure 8 areas of health and HRQOL measures [27]. From SF-8, the Physical Component Summary (PCS-8) score and Mental Component Summary (MCS-8) score were calculated. There are 8 items in the SF-8, and the appropriate SF-36 v2 subscale score is assigned to the category of answers for each item. The subscale of SF-8 is weighted and added by the coefficient for PCS-8 or the coefficient for MCS-8 (coefficient of the regression obtained from Japanese general population data). By adding an intercept of the regression, the scores of PCS-8 and MCS-8 are standardized in the Japanese national data so that they have the same meaning as the summary score of SF-36. The higher the SF-8 score, the better the health condition. The State-Trait Anxiety Inventory (STAI) is a psychological test where the degree of state anxiety and trait anxiety are each scored from a different set of 20 answers [28]. The higher the scores for state and trait anxiety, the greater the anxiety. In this study, we used the standard version of the STAI, in which the past month was considered as the review period.

The study began on a Friday, and subjects performed the assessments on the designated Thursday nights. The mKMI was conducted on the last day of weeks 0, 2, and 4. The SF-8 and STAI were administered on the last day of weeks 0 and 4.

2.6. Statistical Analysis

To estimate the sample size, we referred to previous studies that evaluated the role of dietary nutrients in the health of menopausal women. Hirose et al. investigated the effects of low-dose soy isoflavone aglycone in 90 subjects (each group consisted of 30 subjects) [15], and Terauchi et al. examined the effects of grape seed proanthocyanidin extract in 96 subjects (each group consisted of 31–33 subjects) [17]. In this study, we set the recruitment target for each group to 45 subjects, anticipating that the improvement effect would be difficult to determine because the subjects had a relatively low frequency of hot flushes, which was the primary endpoint.

The indicators representing the characteristics of the subjects at baseline were reported as quartiles and means ± standard deviations (SDs) for continuous variables, and categorical variables were presented as frequencies. The number of hot flushes and the number of sweats were totaled weekly at weeks 0, 1, 2, 3, and 4. The subjects with missing data measurements were excluded from the data at that time. These weekly data were summarized as means and standard errors (SEs) using the Poisson regression model as they were count values. The comparisons between the placebo and CGA groups were conducted using the model, adjusted to use week 0 as the baseline. The objective variable of this model was the number of hot flushes and sweats at week 4 as the primary and secondary endpoints, respectively; however, weeks 1, 2, and 3 were also examined. The amount of change from the baseline VAS assessment was reported as the box plot, and the Wilcoxon rank-sum test was used for comparison.

The mKMI score was also presented as mean ± SE. Lastly, the data from the questionnaire on the various symptoms were compared between groups using the Wilcoxon rank-sum test.

Statistical analyses were conducted using R statistical software and environment (version 4.0.1). All statistical analyses were double-sided. $p < 0.05$ was considered to indicate statistically significant differences.

3. Results

3.1. Number of Hot Flushes and Sweats

At week 0, the weekly mean ± SD of the number of hot flushes in the subjects was 9.4 ± 10.8 in the placebo group ($n = 38$) and 10.2 ± 11.6 in the CGA group ($n = 39$). The weekly average of the entire analytical dataset ($n = 77$) was 9.8 ± 11.1, which indicates that the daily mean for hot flushes at the start of the study was 1.4 ± 1.6. In week 4, the average weekly hot flushes decreased to 4.2 ± 7.6 in the placebo group and 3.7 ± 5.1 in the CGA group. The frequency of sweats also decreased from 6.1 ± 6.8 (week 0) to 2.8 ± 4.5 (week 4) in the placebo group and from 9.6 ± 10.8 (week 0) to 3.5 ± 4.5 (week 4) in the CGA group.

The results of the statistical analysis, after excluding those with missing data at week 4, showed a significant difference in the mean weekly number of hot flushes between the CGA and placebo groups (Figure 2a). The p-values between the groups in terms of the weekly number of hot flushes at week 1, 2, 3 and 4 were 0.433, 0.017, 0.022, and 0.030, and the false discovery rate (FDR)-adjusted p-values, i.e., q-values, were 0.433, 0.040, 0.040, and 0.040, respectively. The results of the subgroup analysis according to menopausal status are shown in the Supplementary Materials (Figure S1). We found no significant differences in the mean weekly number of sweats between groups (Figure 2b). The p-values between the group of the weekly number of sweats at week 1, 2, 3 and 4 were 0.387, 0.823, 0.451, and 0.579, and the q-values were 0.772, 0.823, 0.772, and 0.772, respectively.

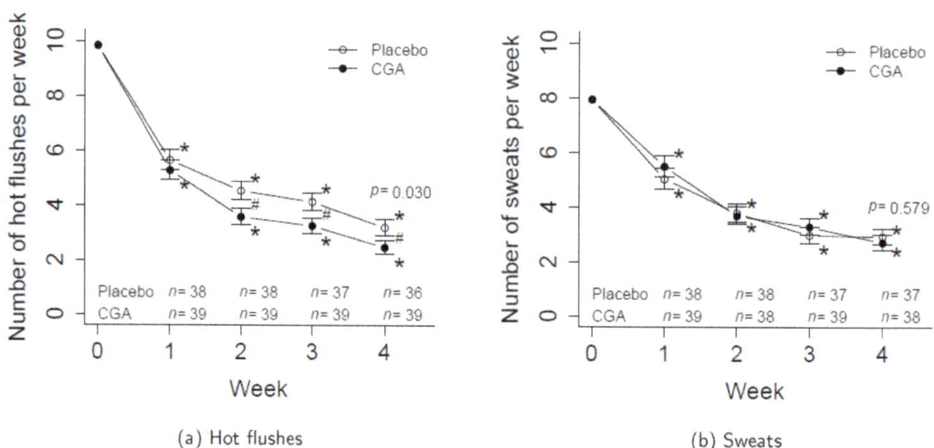

Figure 2. The estimated number of (**a**) hot flushes and (**b**) sweats per week (times/week). Error bars represent standard errors. p-values are the results of the comparison between groups at week 4. [#] represents $p < 0.05$ between groups. * represent $p < 0.05$ within groups. Numbers for weeks 1, 2, 3, and 4 were estimated using the Poisson regression model with adjustment of week 0 number. Participants with missing values were excluded. CGA, chlorogenic acid.

3.2. Severities of Hot Flushes and Sweats (VAS Assessment)

Figure 3 shows the amount of change in the severity of hot flushes and sweats among the subjects compared to week 0 (baseline); the data of three subjects were excluded due to recording errors in either week 0 or 4. At week 4, we found significant differences in the severity of hot flushes during sleep and daytime sweats between the placebo and CGA groups according to the Wilcoxon rank-sum test.

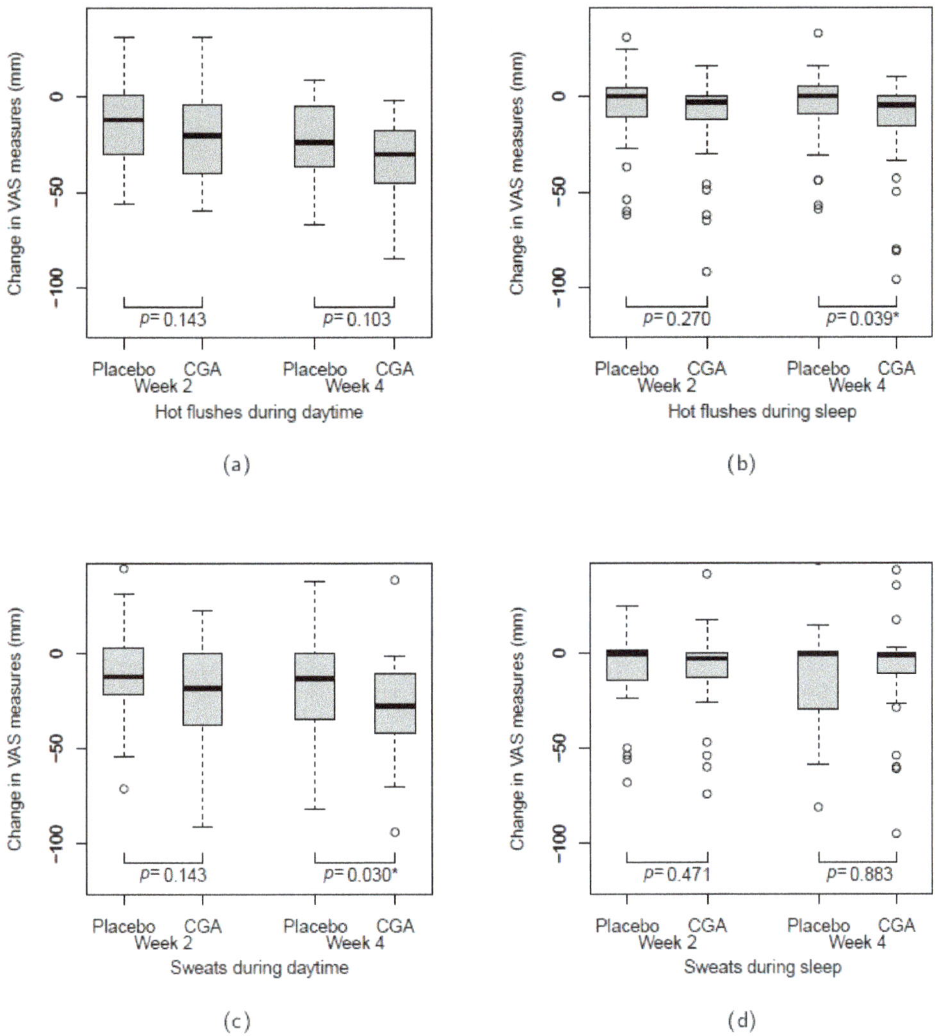

Figure 3. Change in the visual analogue scale (VAS) measurements of hot flushes and sweats compared to week 0: hot flushes during (**a**) the daytime and (**b**) sleep; sweats during (**c**) the daytime and (**d**) sleep. The comparison was assessed using the Wilcoxon rank-sum test. * $p < 0.05$. There was one subject in the chlorogenic acid (CGA) group with missing data for the VAS measurements at week 0. CGA, chlorogenic acid.

3.3. Menopausal Symptoms

Figure 4 shows the results of the mKMI severity grading. The subjects were women experiencing hot flushes who were otherwise healthy, and more than half of the subjects had a mKMI grade of I or II at week 0. Therefore, we conducted a statistical test focusing on the change in the proportion of grade I or II mKMI. Although improvements in menopausal symptoms were observed over time in both the placebo and CGA groups, the improvement was faster in the CGA group than in the placebo group, with a significant difference demonstrated between the groups at week 2 according to the Wilcoxon rank-sum test.

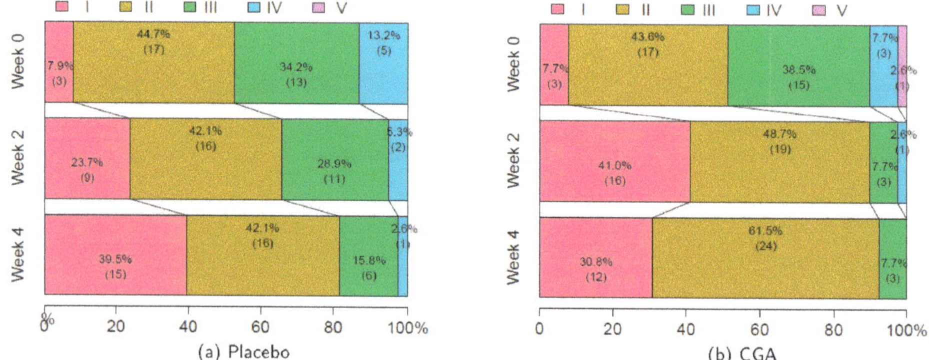

Figure 4. Severity grading based on the modified Kupperman menopausal index scores: (**a**) placebo and (**b**) CGA. Severity grading was assessed according to the following scores: 0–12 is I, 13–22 is II, 23–33 is III, 34–43 is IV, and 44–51 is V. I and II are considered as mild, III as moderate, and IV and V as severe. Comparisons between placebo vs. CGA were $p = 0.906$, $p = 0.016$, and $p = 0.173$ for weeks 0, 2, and 4, respectively, as assessed by a logistic regression model with a threshold score between 22 and 23. Numbers in parentheses indicate frequencies. CGA, chlorogenic acid.

3.4. HRQOL and Anxiety

Table 2 shows the results for the SF-8 and STAI. The PCS-8 scores and MCS-5 scores of the SF-8 significantly increased compared to the initial values in both the placebo and CGA groups, and the state and trait anxiety of STAI decreased in both groups. All findings showed significant improvements compared to the initial values in both groups; however, no significant difference between the two groups was observed.

Table 2. Baseline and change of the Short Form-8 (SF-8) and State-Trait Anxiety Inventory (STAI) scores.

	Baseline		Change from Baseline at Week 4				Difference of Change
	Placebo	CGA	Placebo	*p*-Value	CGA	*p*-Value	*p*-Value
	n = 38	*n* = 39	*n* = 38	Within Groups	*n* = 39	Within Groups	Between Groups
SF-8							
PCS-8 score	43.4 ± 1.0	45.7 ± 1.1	2.6 ± 1.1	0.018 *	2.4 ± 1.1	0.025 *	0.712
MCS-8 score	44.6 ± 1.3	44.7 ± 1.3	4.7 ± 1.1	<0.001 *	3.5 ± 1.2	0.004 *	0.308
STAI							
State anxiety	45.7 ± 1.7	46.5 ± 1.6	−4.8 ± 1.5	0.003 *	−4.5 ± 1.3	0.002 *	0.560
Trait anxiety	50.2 ± 1.9	48.4 ± 1.7	−4.6 ± 1.1	<0.001 *	−4.0 ± 1.1	0.004 *	0.465

Summary statistics are presented as mean ± standard error. *p*-values were derived by Wilcoxon rank-sum test. * $p < 0.05$. CGA, chlorogenic acid; SF, Short Form; PCS, Physical Component Summary; MCS, Mental Component Summary; STAI, State-Trait Anxiety Inventory.

4. Discussion

The hot flushes that afflict menopausal women are classified as vasomotor symptoms that are strongly associated with physical and mental disorders such as insomnia and anxiety, and there are various theories regarding the mechanism underlying their development. We focused on CGAs, dietary nutrients that affect sleep, autonomic nervous system activity, and vascular endothelial function, which are the factors involved in hot flushes, and investigated whether CGAs can reduce the frequency and severity of hot flushes.

When healthy menopausal women were administered 270 mg of CGAs for four weeks, they reported a significant decrease in the number of hot flushes. The subjects in this study were women considered to be healthy except for experiencing an average of 1.4 hot flushes per day, suggesting that the continuous intake of CGAs may further improve the symptoms of relatively infrequent hot flushes. The frequency and severity of vasomotor symptoms, such as hot flushes and sweats, vary depending on the individual, and it is important to address not only the frequency but also the severity of such symptoms. In addition to affecting the QOL during the day, vasomotor symptoms have been shown to cause persistent insomnia and depressive symptoms by interfering with sleep, as suggested by the domino theory [7]. In the present study, we found a significant improvement in the CGA group, namely a reduced severity of sleep hot flushes and daytime sweats (VAS). These results suggest that the continuous intake of CGAs relieves vasomotor symptoms and contributes to improving QOL in menopausal women.

To assess the effectiveness of the continuous intake of CGAs on overall menopausal symptoms, the severity grading was assessed from the total mKMI score, and population changes were observed. The proportion of grade I or II mKMI showed a significant improvement with CGA intake in the second week. This result suggests that some menopausal symptoms may also be ameliorated by the ingestion of CGAs.

The placebo group also showed an improvement in the frequency of hot flashes and sweats from the baseline. Menopausal symptoms have been reported to be affected by psychological status [29]. In this study, taking a test tablet may have raised the expectations of the subjects, and they may have been relieved to find out the effect of taking the tablets from the physical condition records. In the CGA group, in addition to effects of the CGAs, psychological effects also appeared; therefore, we considered that an improvement was experienced earlier in the CGA group compared to the placebo group. These results suggest that a combination of psychotherapy, behavioral therapy, and the intake of CGAs is beneficial.

Why did the continuous intake of CGAs improve menopausal symptoms, especially vasomotor symptoms? There have been many studies on vasomotor symptoms, including the involvement of sleep, autonomic activity, and vascular endothelial function. The domino theory states that vasomotor symptoms cause sleep deprivation and exacerbate psychological symptoms [7], while improving sleep status has been demonstrated to improve vasomotor symptoms [8]. Regarding the autonomic nervous system of menopausal women, Thurston et al. investigated the autonomic nervous activity of peri- and postmenopausal women using power spectrum analysis of heart rate variability [5]. A significant reduction in cardiac vagal control was found to occur during the hot flushes assessed in women's daily activities. Other reports have observed that the sympathovagal balance index value was higher in postmenopausal women [6]. In addition, there is a report that a greater frequency of physiological hot flushes was associated with poorer endothelial function among younger midlife women [4].

CGAs have been reported to have effects on sleep, autonomic nerves, and vascular endothelial function. Ochiai et al. conducted a study in which adult men were given a drink containing 300 mg of CGAs for 2 weeks and reported reduced fatigue upon awakening and significantly improved sleep quality in the CGA group [24]. Between the groups, a significant difference in the sleep efficiency and total nocturnal awaking time in the second half of week 2 was also reported. In a study of nine healthy men and women taking a test beverage containing 600 mg of CGAs for five days, Park et al. found that CGAs shortened sleep latency compared with the control group, as well as enhanced parasympathetic activity, as assessed using heart rate variability during sleep [30].

In addition, Kagawa et al. conducted a study in which 10 healthy men were given a beverage containing 270 mg of CGAs for 4 weeks, and the analysis of heart rate variability showed that CGAs significantly increased parasympathetic nervous activity and decreased sympathetic nervous activity [23]. Based on the above results, the continuous intake of CGAs was expected to enhance parasympathetic nerve activity, improve sleep quality, and reduce daytime fatigue. The continuous intake of CGAs has also been reported to alleviate high blood pressure [31] and improve cutaneous blood flow regulation after cold stress [32].

CGAs are known to have oxygen-scavenging [33], endothelial nitric oxide synthase-activating [34], and nicotinamide adenine dinucleotide phosphate oxidase-inhibiting [35] effects. These effects suggest that CGA improves the bioavailability of nitric oxide in vascular endothelial cells. Therefore, the effects of CGAs on sleep, autonomic nerves, and vascular endothelial function appear to result in improved vasomotor symptoms.

CGAs are widely found in foods, such as coffee beans, apples, pears, tomatoes, potatoes, and eggplants. Of these foods, coffee is rich in CGAs, e.g., single cup of coffee contains 27–121 mg of CGAs [36]. However, coffee also contains caffeine, which exacerbates menopausal symptoms [37]. Therefore, we developed high concentrated CGA tablets, as the form of a tablet may be useful in terms of efficient and convenient intake of CGAs.

Methods for improving menopausal symptoms include HRT and the consumption of medications, such as Japanese traditional herbal medicines (*kampo*), herbs such as black cohosh, and dietary nutrients such as isoflavone, equol, and proanthocyanidin. The present findings show that the continuous intake of CGAs, which are a polyphenol found in coffee and other foods, relieved hot flushes in healthy women. Accordingly, the results of our study show that the consumption of CGAs is a potential method for improving menopausal symptoms.

The limitations of this study include the subject population being limited to those with mild symptoms, no objective assessment of sleep, and lack of measurement of autonomic nervous system activity and vascular function. To further determine the mechanism through which menopausal symptoms of healthy women are improved by CGA intake, a diverse range of subjects should be recruited in future studies, with monitoring of menopausal symptoms and measurement of sleep, autonomic nervous system activity, and vascular function.

5. Conclusions

The continuous intake of CGAs appears to result in improvement in menopausal symptoms, especially hot flushes, in healthy Japanese women.

Supplementary Materials: The following are available online at http://www.mdpi.com/2072-6643/12/12/3757/s1, Figure S1: Subgroup analysis of the estimated number of hot flushes per week (times/week).

Author Contributions: Y.E., study plan and design, study execution, and manuscript preparation; A.S., study plan and design and manuscript preparation; T.Y., statistical analysis and manuscript preparation; R.O., advice on study plan, assistance in study execution, and manuscript preparation; M.T., guidance on manuscript preparation; K.K., supervision of study plan, study execution and manuscript preparation. All authors have read and agreed to the published version of the manuscript.

Funding: This research received no external funding. This work was supported by Kao Corporation (Tokyo, Japan).

Acknowledgments: The authors thank the subjects and physician for their cooperation in this study. They also thank Kao Corporation employees Naoki Yamamoto, Mayumi Otsuka, and Yoko Sugiura for their advice.

Conflicts of Interest: Y.E., A.S., T.Y., R.O. and K.K. are employees of Kao Corporation (Tokyo, Japan).

References

1. Freedman, R.R. Hot flashes: Behavioral treatments, mechanisms, and relation to sleep. *Am. J. Med.* **2005**, *118*, 124–130. [CrossRef] [PubMed]
2. Mittelman-Smith, M.A.; Williams, H.; Krajewski-Hall, S.J.; McMullen, N.Y.; Rance, N.E. Role for kisspeptin/neurokinin B/dynorphin (KNDy) neurons in cutaneous vasodilatation and the estrogen modulation of body temperature. *Proc. Natl. Acad. Sci. USA* **2012**, *109*, 19846–19851. [CrossRef] [PubMed]
3. Hay, D.L.; Poyner, D.R. Calcitonin gene-related peptide, adrenomedullin and flushing. *Maturitas* **2009**, *64*, 104–108. [CrossRef] [PubMed]

4. Thurston, R.C.; Chang, Y.; Barinas-mitchell, E.; Jennings, J.R.; von Känel, R.; Landsittel, D.P.; Matthews, K.A. Physiologically assessed hot flashes and endothelial function among midlife women. *Menopause* **2018**, *25*, 1354–1361. [CrossRef]

5. Thurston, R.C.; Christie, I.C.; Matthews, K.A. Hot flashes and cardiac vagal control during women's daily lives. *Menopause* **2012**, *19*, 406–412. [CrossRef]

6. Moodithaya, S.S.; Avadhany, S.T. Comparison of cardiac autonomic activity between pre and post menopausal women using heart rate variability. *Indian J. Physiol. Pharmacol.* **2009**, *53*, 227–234.

7. Eichling, P.S.; Sahni, J. Menopause related sleep disorders. *J. Clin. Sleep Med.* **2005**, *1*, 291–300. [CrossRef]

8. Joffe, H.; Petrillo, L.; Viguera, A.; Koukopoulos, A.; Silver-Heilman, K.; Farrell, A.; Yu, G.; Silver, M.; Cohen, L.S. Eszopiclone improves insomnia and depressive and anxious symptoms in perimenopausal and postmenopausal women with hot flashes: A randomized, double-blinded, placebo-controlled crossover trial. *Am. J. Obstet. Gynecol.* **2010**, *202*, 171.e1–171.e11. [CrossRef]

9. Rossouw, J.E.; Anderson, G.L.; Prentice, R.L.; LaCroix, A.Z.; Kooperberg, C.; Stefanick, M.L.; Jackson, R.D.; Beresford, S.A.; Howard, B.V.; Johnson, K.C.; et al. Risks and benefits of estrogen plus progestin in healthy postmenopausal women: Principal results From the Women's Health Initiative randomized controlled trial. *JAMA* **2002**, *288*, 321–333. [CrossRef]

10. Sturdee, D.W.; Pines, A.; Archer, D.F.; Baber, R.J.; Barlow, D.; Birkhäuser, M.H.; Brincat, M.; Cardozo, L.; de Villiers, T.J.; Gambacciani, A.A.; et al. Updated IMS recommendations on postmenopausal hormone therapy and preventive strategies for midlife health. *Climacteric* **2011**, *14*, 302–320. [CrossRef]

11. Chen, J.T.; Shiraki, M. Menopausal hot flash and calciotonin gene-related peptide; Effect of Keishi-bukuryo-gan, a kampo medicine, related to plasma calciotonin gene-related peptide level. *Maturitas* **2003**, *45*, 199–204. [CrossRef]

12. Yasui, T.; Matsui, S.; Yamamoto, S.; Uemura, H.; Tsuchiya, N.; Noguchi, M.; Yuzurihara, M.; Kase, Y.; Irahara, M. Effects of Japanese traditional medicines on circulating cytokine levels in women with hot flashes. *Menopause* **2011**, *18*, 85–92. [CrossRef] [PubMed]

13. Jiang, K.; Jin, Y.; Huang, L.; Feng, S.; Hou, X.; Du, B.; Zheng, J.; Li, L. Black cohosh improves objective sleep in postmenopausal women with sleep disturbance. *Climacteric* **2015**, *18*, 559–567. [CrossRef] [PubMed]

14. Chen, L.R.; Ko, N.Y.; Chen, K.H. Isoflavone supplements for menopausal women: A systematic review. *Nutrients* **2019**, *11*, 2649. [CrossRef]

15. Hirose, A.; Terauchi, M.; Akiyoshi, M.; Owa, Y.; Kato, K.; Kubota, T. Low-dose isoflavone aglycone alleviates psychological symptoms of menopause in Japanese women: A randomized, double-blind, placebo-controlled study. *Arch. Gynecol. Obstet.* **2016**, *293*, 609–615. [CrossRef]

16. Aso, T.; Uchiyama, S.; Matsumura, Y.; Taguchi, M.; Nozaki, M.; Takamatsu, K.; Ishizuka, B.; Kubota, T.; Mizunuma, H.; Ohta, H. A natural S-equol supplement alleviates hot flushes and other menopausal symptoms in equol nonproducing postmenopausal Japanese women. *J. Womens Health* **2012**, *21*, 92–100. [CrossRef]

17. Terauchi, M.; Horiguchi, N.; Kajiyama, A.; Akiyoshi, M.; Owa, Y.; Kato, K.; Kubota, T. Effects of grape seed proanthocyanidin extract on menopausal symptoms, body composition, and cardiovascular parameters in middle-aged women: A randomized, double-blind, placebo-controlled pilot study. *Menopause* **2014**, *21*, 990–996. [CrossRef]

18. El Khoudary, S.R.; Greendale, G.; Crawford, S.L.; Avis, N.E.; Brooks, M.M.; Thurston, R.C.; Karvonen-Gutierrez, C.; Waetjen, L.E.; Matthews, K. The menopause transition and women's health at midlife: A progress report from the Study of Women's Health across the Nation (SWAN). *Menopause* **2019**, *26*, 1213–1227. [CrossRef]

19. Mori, H.; Tanaka, T.; Shima, H.; Kuniyasu, T.; Takahashi, M. Inhibitory effect of chlorogenic acid on methylazoxymethanol acetate-induced carcinogenesis in large intestine and liver of hamsters. *Cancer Lett.* **1986**, *30*, 49–54. [CrossRef]

20. Natella, F.; Nardini, M.; Giannetti, I.; Dattilo, C.; Scaccini, C. Coffee drinking influences plasma antioxidant capacity in humans. *J. Agric. Food Chem.* **2002**, *50*, 6211–6216. [CrossRef]

21. Suzuki, A.; Yamamoto, N.; Jokura, H.; Yamamoto, M.; Fujii, A.; Tokimitsu, I.; Saito, I. Chlorogenic acid attenuates hypertension and improves endothelial function in spontaneously hypertensive rats. *J. Hypertens.* **2006**, *24*, 1065–1073. [CrossRef] [PubMed]

22. Soga, S.; Ota, N.; Shimotoyodome, A. Stimulation of postpransial fat utilization in healthy humans by daily consumption of chlorogenic acids. *Biosci. Biotechnol. Biochem.* **2013**, *77*, 1633–1636. [CrossRef] [PubMed]

23. Kagawa, D.; Fujii, A.; Ohtsuka, M.; Murase, T. Ingestion of coffee polyphenols suppresses deterioration of skin barrier function after barrier disruption, concomitant with the modulation of autonomic nervous system activity in healthy subjects. *Biosci. Biotechnol. Biochem.* **2018**, *82*, 879–884. [CrossRef]

24. Ochiai, R.; Tomonobu, K.; Ikushima, I. Effect of chlorogenic acids on fatigue and sleep in healthy males: A randomized, double-blind, placebo-controlled, crossover study. *Food Sci. Nutr.* **2018**, *6*, 2530–2536. [CrossRef] [PubMed]

25. Abe, T.; Yamaya, Y.; Suzuki, M.; Moritsuka, T. Statistical clustering of women complaining of climacteric syndrome by cluster analysis. *Acta Obstet. Gynaecol. Jpn.* **1979**, *31*, 607–614.

26. Kupperman, H.S.; Blatt, M.H.; Wiesbader, H.; Filler, W. Comparative clinical evaluation of estrogenic preparations by the menopausal and amenorrheal indices. *J. Clin. Endocrinol. Metab.* **1953**, *13*, 688–703. [CrossRef]

27. Fukuhara, S.; Suzukamo, Y. *Manual of SF-8 Japanese Version*; Institute for Health Outcomes & Process Evaluation Research: Kyoto, Japan, 2004.

28. Spielberger, C.D.; Gorsuch, R.L.; Lushene, R.; Vagg, P.R.; Jacobs, G.A. *Manual for the State-Trait Anxiety Inventory*; Consulting Psychologists Press: Kansas City, MO, USA, 1983.

29. Hunter, M.; Battersby, R.; Whitehead, M. Relationships between psychological symptoms, somatic complaints and menopausal status. *Maturitas* **1986**, *8*, 217–228. [CrossRef]

30. Park, I.; Ochiai, R.; Ogata, H.; Kayaba, M.; Hari, S.; Hibi, M.; Katsuragi, Y.; Satoh, M.; Tokuyama, K. Effects of subacute ingestion of chlorogenic acids on sleep architecture and energy metabolism through activity of the autonomic nervous system: A randomised, placebo-controlled, double-blinded cross-over trial. *Br. J. Nutr.* **2017**, *117*, 979–984. [CrossRef]

31. Yamaguchi, T.; Chikama, A.; Mori, K.; Watanabe, T.; Shioya, Y.; Katsuragi, Y.; Tokimitsu, I. Hydroxyhydroquinone-free coffee: A double-blind, randomized controlled dose-response study of blood pressure. *Nutr. Metab. Cardiovasc. Dis.* **2008**, *18*, 408–414. [CrossRef] [PubMed]

32. Ueda, S.; Tanahashi, M.; Higaki, Y.; Iwata, K.; Sugiyama, Y. Ingestion of coffee polyphenols improves a scaly skin surface and the recovery rate of skin temperature after cold stress: A randomized, controlled trial. *J. Nutr. Sci. Vitaminol.* **2017**, *63*, 291–297. [CrossRef]

33. Suzuki, A.; Yamamoto, M.; Jokura, H.; Fujii, A.; Tokimitsu, I.; Hase, T.; Saito, I. Ferulic acid restores endothelium-dependent vasodilation in aortas of spontaneously hypertensive rats. *Am. J. Hypertens.* **2007**, *20*, 508–513. [CrossRef] [PubMed]

34. Hou, Y.; Yang, J.; Zhao, G.; Yuan, Y. Ferulic acid inhibits endothelial cell proliferation through NO down-regulating ERK1/2 pathway. *J. Cell Biochem.* **2004**, *93*, 1203–1209. [CrossRef] [PubMed]

35. Suzuki, A.; Fujii, A.; Yamamoto, N.; Yamamoto, M.; Ohminami, H.; Kameyama, A.; Shibuya, Y.; Nishizawa, Y.; Tokimitsu, I.; Saito, I. Improvement of hypertension and vascular dysfunction by hydroxyhydroquinone-free coffee in a genetic model of hypertension. *FEBS Lett.* **2006**, *580*, 2317–2322. [CrossRef] [PubMed]

36. Mills, C.E.; Oruna-Concha, M.J.; Mottram, D.S.; Gibson, G.R.; Spencer, J.P.E. The effect of processing on chlorogenic acid content of commercially available coffee. *Food Chem.* **2013**, *141*, 3335–3340. [CrossRef]

37. Faubion, S.S.; Sood, R.; Thielen, J.M.; Shuster, L.T. Caffeine and menopausal symptoms: What is the association? *Menopause* **2015**, *22*, 155–158. [CrossRef]

Publisher's Note: MDPI stays neutral with regard to jurisdictional claims in published maps and institutional affiliations.

© 2020 by the authors. Licensee MDPI, Basel, Switzerland. This article is an open access article distributed under the terms and conditions of the Creative Commons Attribution (CC BY) license (http://creativecommons.org/licenses/by/4.0/).

Review

The Diverse Efficacy of Food-Derived Proanthocyanidins for Middle-Aged and Elderly Women

Toru Izumi [1,*] and Masakazu Terauchi [2]

[1] Department of Production and Quality Control, Kikkoman Nutricare Japan Incorporation, Nihonbashikoamicho 3-11, Chuo, Tokyo 103-0016, Japan

[2] Department of Women's Health, Tokyo Medical and Dental University, Yushima 1-5-45, Bunkyo, Tokyo 113-8510, Japan; teragyne@tmd.ac.jp

* Correspondence: tizumi@mail.kikkoman.co.jp; Tel.: +81-3-5521-5138; Fax: +81-3-3660-9222

Received: 29 October 2020; Accepted: 12 December 2020; Published: 15 December 2020

Abstract: Middle-aged and elderly women are affected by various symptoms and diseases induced by estrogen deficiency. Proanthocyanidins, widely present in many kinds of fruits and berries, have many beneficial effects, such as antioxidative, anti-inflammatory, and antimicrobial activities. We researched the effects of proanthocyanidins for middle-aged and elderly women, finding that it has been revealed in many clinical trials and cohort studies that proanthocyanidins contribute to the prevention of cardiovascular disease, hypertension, obesity, cancer, osteoporosis, and urinary tract infection, as well as the improvement of menopausal symptoms, renal function, and skin damage. Thus, proanthocyanidins can be considered one of the potent representatives of complementary alternative therapy.

Keywords: proanthocyanidin; middle-aged; women; menopause; cardiovascular disease; grape seed; cranberry

---

1. Introduction

Menopause results from reduced secretion of the ovarian hormones estrogen and progesterone, which takes place as the finite store of ovarian follicles is depleted. Women in the menopausal transition and postmenopausal period are affected by vasomotor symptoms, urogenital atrophy, sexual dysfunction, somatic symptoms, cognitive difficulty, sleep disturbance, and psychological problems. Some of these effects such as vasomotor symptoms are closely associated with estrogen deficiency, but the exact mechanisms underlying the other symptoms are not fully understood [1]. Postmenopausal women are also at increased risk for cardiovascular morbidity [2] as a net effect of central obesity [3], dyslipidemia [4], hypertension [5], and diabetes [6] partly induced by estrogen deficiency.

Flavonoids are a class of polyphenolic compounds with significant human health benefits [7]. Some of the flavonoids such as the flavan-3-ols catechin and epicatechin polymerize to form tannins. Tannins are plant secondary metabolites that can be hydrolysable or condensed. The condensed tannins are also known as proanthocyanidins [8,9]. They are widely present in the plant kingdom, for example, in fruits, berries, nuts, seeds, and bark of pine trees [10–12]. Proanthocyanidins are oligomers or polymers of flavan-3-ols, where the monomeric units are linked mainly by C-4 to C-8 bonds, although less frequently C-4 to C-6 linkages can also be found. These types of linkages lead to the formation of the so-called B-type proanthocyanidins. A-type proanthocyanidins, on the other hand, are characterized by an additional bond between C-2 $\rightarrow$ C-7 of the basic flavan-3-ol units [13]. Proanthocyanidins are composed of different flavan-3-ol subunits, known as proanthocyanidin monomers or catechins. The most common monomeric units are the diastereomers of (epi)catechin, (epi)afzelechin, and (epi)gallocatechin (Figure 1).

Proanthocyanidin A-type

Proanthocyanidin B-type

Compound	⋀OH	R₁	R₂
Afzelechin		H	H
Catechin	►◄OH	OH	H
Gallocatechin		OH	OH
Epiafzelechin		H	H
Epicatechin	⋯OH	OH	H
Epigallocatechin		OH	OH

Figure 1. Basic chemical features of A-and B-type proanthocyanidinsand their monomeric units.

In recent years, considerable attention has been paid to proanthocyanidins due to the potential beneficial effects on human health, including antioxidative, cardio-protective anti-inflammatory, and anticancer properties [13–16]. Some reviews have also been published on the effects of proanthocyanidins [9,13], however, there still seems to be no reviews on the effects for middle-aged and elderly women. This review therefore aims to highlight aspects of the effects of food-derived proanthocyanidins for middle-aged and elderly women's health.

2. Search Strategy and Study Selection

A literature search of all English language studies published was performed using PubMed (http://www.ncbi.nlm.nih.gov/pubmed) with the addition of other scientific papers of relevance found in web sources or in previously published reviews. The search terms and strategy used for the study selection were (proanthocyanidins OR procyanidins OR flavan 3-ol polymers) AND (women OR female) AND clinical trials. Human studies were used as further criteria for the literature search. The search was limited to the last 20 years of publication. We conducted the literature search in the scientific databases and assessed and verified the eligibility of the studies on the basis of the title and abstract. Inclusion criteria: (i) randomized controlled trials (RCT), randomized cross-over trials, quasi-randomized controlled trial (qRCT); (ii) long-term (over 6 months) clinical trials; (iii) prospective, cohort, and case-control studies. The exclusion criteria were (i) studies performed in only male or only under 40-aged women; (ii) studies analyzing the relationship between other polyphenols (not proanthocyanidins) and the occurrence of disease; (iii) studies performed in in vitro or in animal models; (iv) studies reporting data on the disease that is not related to middle-aged and elderly women; and (v) published articles in a language different from English and with no accessible translation.

3. Results

Table 1 is a summary showing current evidence of studies regarding the effects of proanthocyanidins for middle-aged and elderly women. We report the detail results in each field as follows: menopausal disorders, cancer, hypertension, cardiovascular disease, obesity, osteoporosis, urinary tract infection, renal function, and skin damage. In addition, we unfortunately could not find any reports on the effects of the disease that increase in the postmenopausal period, such as osteoarthritis, dyslipidemia, cognitive impairment, overactive bladder, urinary incontinence, vulvovaginal atrophy.

Table 1. A summary of current evidence of studies regarding the effects of proanthocyanidins for middle-aged and elderly women.

Studies (Ref. No.)	Study Design	Period	Age	Contents	Main Results
Menopausal Disorders (Including Mental Symptoms)					
Terauchi [17]	RCT	8 weeks	40–60	grape seed extract proanthocyanidins 100 mg/day or 200 mg/day	improvement of HADS-anxiety, HADS-depression and AIS score
Kohama [18]	RCT	12 weeks	42–58	pine bark extract proanthocyanidins 30 mg/day	improvement of vasomotor and insomnia/sleep problem symptoms and Kupperman's index
Yang [19]	RCT	6 months	45–55	pine bark extract proanthocyanidins 200 mg/day	improvement of climacteric symptoms
Chang [20]	Prospective cohort	10 years	65-	diet-derived proanthocyanidins	lowering of depression risk
Cancer					
Touvier [21]	Prospective cohort	14 years	55-	diet-derived proanthocyanidins	lowering of breast cancer risks
Rossi [22]	case-control	15 years	60–61 (median)	diet-derived proanthocyanidins	lowering of endometrial cancer risk
Culter [23]	Prospective cohort	19 years	55–69	diet-derived proanthocyanidins	decreasing of lung cancer incidence
Hypertension					
Terauchi [17]	RCT	8 weeks	40–60	grape seed extract proanthocyanidins 100 mg/day or 200 mg/day	reducing of SBP and DBP
Odai [24]	RCT	12 weeks	40–64	grape seed extract proanthocyanidins 200 mg/day or 400 mg/day	reducing of SBP and DBP
Lajous [25]	Prospective cohort	16 years	45–58	diet-derived proanthocyanidins	lowering of hypertension rate
Cardiovascular disease					
Odai [24]	RCT	12 weeks	40–64	grape seed extract proanthocyanidins 200 mg/day or 400 mg/day	Improvement of stiffness parameter, distensibility, Eincand PWV
Zhao [26]	RCT	4 weeks	42–53	apple proanthocyanidins	decreasing of oxLDL-b2GPI
Mink [27]	Prospective cohort	13 years	55–69	diet-derived proanthocyanidins	inverse association with coronary heart disease mortality
McCullough [28]	Prospective cohort	7 years	68.9 ± 6.2	diet-derived proanthocyanidins	reduction in cardiovascular disease risk
Jennings [29]	Prospective cohort	12 years	53 (median)	diet-derived proanthocyanidins	lowering of fat mass ratio (FMR)

Table 1. *Cont.*

Studies (Ref. No.)	Study Design	Period	Age	Contents	Main Results
Obesity					
Tresserra-Rimbau [30]	Prospective cohort	4 years	60–75	diet-derived proanthocyanidins	inverse association with overweight and obese
Kim [31]	Prospective cohort	15 years	45.0 ± 0.2	diet-derived proanthocyanidins	inverse association with abdominal obesity
Osteoporosis					
Panahande [32]	RCT	12 weeks	50–65	pine bark extract proanthocyanidins 250 mg/day	increase of P1NP and BAP levels, decrease of CTx1 levels
Zhang [33]	Prospective cohort	12 weeks	56–63	diet-derived proanthocyanidins	Increase of bone mineral density at the whole body, femoral neck and lumbar spine
Urinary Tract Infection (UTI)					
Takahashi [34]	RCT	6 months	50-	cranberry proanthocyanidins 40 mg/day (cranberry juice)	prevention of the recurrence of UTI
Maki [35]	RCT	24 weeks	40–41 (average)	cranberry proanthocyanidins 41 mg/day (cranberry juice)	lowered the number of clinical UTI episodes
Vostalova [36]	RCT	6 months	18–75	cranberry powder 500 mg/day (proanthocyanidins 0.56%)	reduction of the risk of symptomatic UTI
Renal Function					
Ivey [37]	Prospective cohort	5 years	80 ± 3	diet-derived proanthocyanidins	improvement of renal function and reduction of risk of chronic kidney disease and renal disease associated events
Skin Damage					
Yamakoshi [38]	single-armed	12 months	34–58	grape seed extract proanthocyanidins 160 mg/d	improvement of chloasma and decreasing melanin index
Evangeline [39]	RCT	8 weeks	18–60	cranberry proanthocyanidins 24 mg/day	decreasing the degree of pigmentation in the malar regions and improvement of the melasma area and severity index
Furumura [40]	RCT	12 weeks	31–59	pine bark extract proanthocyanidins 100 mg/day	improvement of scores for solar lentigines, mottled pigmentation, roughness, wrinkles, and swelling

HADS: Hospital Anxiety and Depression Scale; SBP: systolic blood pressure; DBP: diastolic blood pressure; BAP: bone alkaline phosphatase; P1NP: procollagen type 1 amino-terminal propeptide; CTx1: C-terminal telopeptide of type I collagen.

3.1. Menopausal Disorders

In a randomized, double-blind, placebo-controlled study in Japanese women aged 40 to 60 years who received grape seed proanthocyanidins for 8 weeks, Terauchi et al. evaluated the effect for menopausal symptoms. The mean physical symptom score for the nine items in the physical health domain of the Menopausal Health-Related Quality of Life (MHR-QOL) questionnaire significantly improved in the high-dose (200 mg proanthocyanidins/day) group after 8 weeks of treatment ($p < 0.05$). The mean score for hot flashes similarly improved in the high-dose group after 8 weeks ($p < 0.05$). In the case of changes in psychological symptoms, the mean (Hospital Anxiety and Depression Scale (HADS)-Depression subscale score did not change significantly in any group, whereas the mean HADS-Anxiety subscale score improved in both low-dose (100 mg proanthocyanidins/day) and high-dose groups after 4 weeks of treatment. The change in HADS-Anxiety subscale score from baseline to 8 weeks was significantly higher in the high-dose group than in the placebo group ($p < 0.05$). The mean Athens Insomnia Scale also improved in the high-dose group after 8 weeks [17]. Pine bark extract proanthocyanidins have been evaluated for improvement of climacteric symptoms in two separate clinical studies. A relatively low daily dosage of pine bark extract proanthocyanidins were found to be especially effective for improving vasomotor and insomnia/sleep problem symptoms, which were significantly better after 4 and 12 weeks than with placebo in perimenopausal Taiwanese women ($p < 0.05$). Total Kupperman's index for perimenopausal symptom severity score decreased significantly compared to placebo after 12 weeks of treatment ($p < 0.05$). Symptom score was also significantly better after 4 weeks of treatment with active as compared to placebo [18]. Yang et al. reported that in 155 peri-menopausal Taiwanese women, many climacteric symptoms of the Women's Health Questionnaire significantly improved compared to start of treatment ($p < 0.001$) after 6 months as follows: somatic problems, depression, vasomotoric problems, memory/concentration, attractiveness, anxiety, sexual behavior, sleep, and menstrual problems [19]. In a large prospective female cohort study with 10-year of follow up, Chang et al. reported that in flavonoid-rich food-based analyses, the hazard ratio (HR) was 0.82 (95% confidence interval (CI): 0.74–0.91) among participants who consumed up to two servings of citrus fruit or juices per day compared with below 1 serving per week. In the participants alone, total flavonoids, polymers, and proanthocyanidin intakes showed significantly (9–12%) lower depression risk. In analyses among late-life participants (aged up to 65 years at baseline or during follow-up), for whom they could incorporate depressive symptoms into the outcome definition, the researchers found that higher intakes of all flavonoid subclasses except for flavan-3-ols were associated with significantly lower depression risk; proanthocyanidins showed the strongest associations (HR: 0.83; 95% CI: 0.90) [20].

3.2. Cancer

In the SU.VI.MAX prospective cohort study followed from 1994 to 2007 in France, Touvier et al. revealed that in 55 older-aged women who were non-to-low alcohol drinkers, the intake of proanthocyanidins was associated with significantly decreased breast cancer risk ($p < 0.05$) [21]. Rossi et al. analyzed data from an Italian case–control study including 454 incidents of histologically confirmed endometrial cancers and 908 hospital-based controls to examine the relationship between dietary flavonoids and endometrial cancer. Women in the highest quartile category of proanthocyanidins with trimers vs. the first three quartile categories had an odds ratio for endometrial cancer of 0.66 (95% CI = 0.48–0.89). High consumption of selected proanthocyanidins may reduce endometrial cancer risk [22]. In another study wherein participants were 34,708 postmenopausal women in the Iowa Women's Health Study who completed a food frequency questionnaire and were followed for cancer occurrence from 1986 through 2004, after multivariable adjustment, lung cancer incidence was inversely associated with intakes of proanthocyanidins (HR = 0.75; 95% CI: 0.57–0.97, which resulted in the highest vs. lowest quintile). Among current and past smokers, those with intakes in the highest quintile for proanthocyanidins (HR 5 = 0.66; 95% CI: 0.49–0.89) had significantly lower lung cancer incidence than those in the lowest quintile [23].

3.3. Hypertension

In a randomized, double-blind, placebo-controlled study in Japanese women aged 40 to 60 years who received grape seed proanthocyanidins for 8 weeks, the mean systolic and diastolic blood pressure were significantly reduced in both low-dose (100 mg proanthocyanidin per day) and high-dose (200 mg proanthocyanidin per day) groups after 4 weeks of treatment. The change in diastolic blood pressure (DBP) from baseline to 8 weeks was significantly higher in the low-dose and high-dose groups than in the placebo group ($p < 0.05$) [17]. Additionally, Odai et al. showed in a randomized, double-blind study of 21 non-smoking middle-aged women (including men), the oral intake of grape seed proanthocyanidins significantly reduced the mean systolic blood pressure (SBP) and DBP by 13.1 and 6.5 mmHg, respectively, in the high-dose (400 mg proanthocyanidin per day) group after 12 weeks of intervention [24]. A cohort study reported the anti-hypertensive effect of proanthocyanidins on middle-aged and elderly women. In a prospective cohort of 40,574 disease-free French women who responded to a validated dietary questionnaire, there was a 9% lower rate of hypertension for women in the highest category of proanthocyanidin intake than for women in the lowest category of intake of proanthocyanidin (HR: 0.91 (95% CI: 0.85–0.97; p-trend = 0.0051)) [25].

3.4. Cardiovascular Disease

Odai et al. conducted a randomized, double-blind, placebo-controlled study on 6 men and 24 women aged 40–64 years old who received tablets containing either low-dose (200 mg/day) or high-dose (400 mg/day) grape seed extract proanhocyanidin, or placebo, for 12 weeks. In an ad hoc analysis of non-smoking participants ($n = 21$), stiffness parameter, distensibility, incremental elastic modulus (Einc), and pulse wave velocity (PWV) also significantly improved in the high-dose group after 12 weeks. Changes in Einc and PWV from baseline to 12 weeks were significantly greater in the high-dose group than in the placebo group (Einc, $p = 0.023$; PWV, $p = 0.03$) [24]. Zhao et al. performed a clinical trial to evaluate the effect of apple or apple extract polyphenol on a blood parameter, oxidized low-density lipoprotein/beta2-glycoprotein I complex (oxLDL-b2GPI), related to atherosclerosis in postmenopausal American women. Consumption of apples or apple polyphenols produced a statistically significant decrease in oxLDL-b2GPI, while placebo did not produce a significant change. The mean decrease was larger for the apples than for the apple polyphenols. The apple extract polyphenol and the apples both had their largest polyphenol contribution from proanthocyanidins.

Therefore, eating one apple containing mainly proanthocyanidins per day had a substantial effect on a blood parameter related to atherosclerosis [26]. In a prospective study in postmenopausal women in the Iowa Women's Health Study who were free of cardiovascular disease (CVD) and had complete food-frequency questionnaire information at baseline, Mink et al. evaluated the association between flavonoid intake and CVD mortality. As a result, the intake of proanthocyanidins was significantly inversely associated with coronary heart disease (CHD) mortality in models after adjustment for age and energy ($p < 0.05$), and for total CVD mortality, a significant inverse association after adjustment for age and energy intake was observed for intake of proanthocyanidins ($p < 0.01$) [27]. McCullough et al. also reported a large U.S. cohort study for the association between flavonoid intake and CVD mortality among U.S. participants. In 60,289 women in the Cancer Prevention Study II Nutrition Cohort with a mean age of 69 years, proanthocyanidins consumption was associated with the significant great reduction in CVD risk for the highest intake quintile compared with the lowest quintile ($p = 0.01$) [28]. Jennings et al. showed the effect of proanthocyanidins for fat mass ratio (FMR) in a study of healthy middle-aged female twins. In monozygotic, intake-discordant twin pairs, twins with higher intakes of proanthocyanidins ($p = 0.01$) had a significantly lower FMR than that of their cotwins with within-pair differences of 3–4% [29]. Some meta-analysis for the association between the flavonoid intake and CVD or CHD risk has been reported so far [41–45], but the stratified analysis for middle-aged and elderly women have not yet been performed. Therefore, we failed to evaluate the efficacy of proanthocyanidins to women for those meta-analysis.

3.5. Obesity

Tresserra-Rimbau et al. assessed whether high intakes of some classes of polyphenols including proanthocyanidins were associated with type 2 diabetes (T2D) in a population with metabolic syndrome and how these associations depend on body mass index (BMI) and sex. This baseline cross-sectional analysis included 6633 participants from the PREDIMED-Plus trial. They found in a case of proanthocyanidins for elderly Spanish women that there were significant and linear inverse associations in comparing extreme quartiles for overweight and obesity, recognized as important risk factors for T2D [30]. In another cohort study in Korea, Kim et al. investigated the association between dietary flavonoid intake and the prevalence of obesity using body mass index (BMI), waist circumference, and percent body fat (%BF) among elderly Korean females. A higher total intake of flavonoids was associated with a lower prevalence of obesity in women, on the basis of %BF (odds ratio (95% CI) = 0.82 (0.71–0.94)), and abdominal obesity (0.81 (0.71–0.92)). The intake of proanthocyanidins (0.81 (0.71–0.92)) was inversely associated with abdominal obesity, and a higher intake of proanthocyanidins (0.85 (0.75–0.98)) was associated with a lower prevalence of obesity with respect to %BF in women [31]. Some meta-analyses for the association between flavonoid intake and T2D or overweight risk have been reported thus far [46,47], but the stratified analysis for middle-aged and elderly women have not yet been performed. Therefore, we failed to evaluate the efficacy of proanthocyanidins to women for those meta-analyses.

3.6. Osteoporosis

Panahande et al. reported the effects of pine bark extract procyanidin on bone remodeling in postmenopausal osteopenic women in a randomized, double-blinded, controlled clinical trial for 12 weeks. After the 12-week intervention, that is, pine bark extract procyanidin supplementation, a significant increase in bone alkaline phosphatase (BAP) and procollagen type 1 amino-terminal propeptide (P1NP) levels, and a significant decrease in C-terminal telopeptide of type I collagen (CTx1) were observed. Compared with the control group, pine bark extract procyanidin supplementation resulted in a significant increase in P1NP levels ($p < 0.05$), BAP levels ($p < 0.01$), and BAP/CTx1 ratio ($p < 0.01$), and a significant decrease in CTx1 levels ($p < 0.01$) [32]. Additionally, a large cross-sectional study examined the associations of dietary intake of total flavonoids and their subtypes with bone density in elderly Chinese women. Zhang et al. showed that after adjusting for covariates, women who consumed higher proanthocyanidins tended to have greater bone mineral density in the whole body, femoral neck, and lumbar spine (all trend $p < 0.05$). Women in the highest (vs. the lowest) quartile of proanthocyanidin intake had 0.014–0.016 g/cm^2 greater BMD in the whole body, femoral neck, and lumbar spine [33].

3.7. Urinary Tract Infection

Takahashi et al. performed a randomized, placebo-controlled, double-blind study to examine the rate of relapse in Japanese patients with urinary tract infection (UTI) who suffered from multiple relapses when using cranberry juice containing 40 mg of proanthocyanidin for 24 weeks. In the group of females aged 50 years or more, there was a significant difference in the rate of relapse of UTI between groups cranberry juice and placebo ($p < 0.05$) [34]. Aside from this study, Maki et al. reported another clinical study for UTI recurrence using cranberry juice. In a randomized, double-blind, placebo-controlled, multicenter clinical trial, women with a history of a recent UTI were assigned to consume one 240 mL serving of cranberry beverage or a placebo beverage for 24 weeks. The annualized UTI incidence density was significantly reduced in the cranberry compared with the placebo group (incidence rate ratio: 0.62; 95% CI: 0.42–0.92; $p < 0.05$). The consumption of 41.1 mg cranberry proanthocyanidin in a beverage lowered the number of clinical UTI episodes in women with a recent history of UTI [35]. Vostalova et al. showed that the study tested whether whole cranberry fruit powder (proanthocyanidin content 0.56%) could prevent recurrent UTI in 182 women with two or more UTI episodes. Participants were randomized to a cranberry ($n = 89$) or a placebo group ($n = 93$) and received daily 500 mg of cranberry for 6 months. The number of UTI diagnoses was counted. The intent-to-treat analyses showed that in the cranberry group, the UTIs were significantly fewer (10.8% vs. 25.8%, $p = 0.04$, with an age-standardized 12-month UTI history ($p < 0.05$) [36].

3.8. Renal Function

Ivey et al. performed a cohort study to determine the association of habitual proanthocyanidin intake with renal function and the risk of clinical renal outcomes in a population of elderly Caucasian women. Compared to participants with low consumption, participants in the highest tertile of proanthocyanidin intake had a 9% lower cystatin C concentration ($p < 0.001$). High proanthocyanidin consumers were at 50% lower risk of moderate chronic kidney insufficiency, and 65% lower risk of experiencing a 5-year renal disease event ($p < 0.05$). Proanthocyanidin intake was associated with improved renal function and reduced risk of chronic kidney disease and renal disease-associated events [37].

3.9. Skin Damage

Yamakoshi et al. investigated the reducing effect of grape seed extract proanthocyanidin (GSEP) on chloasma for 12 middle-aged Japanese women in a one-year open design study. The first 6 months of GSEP intake improved or slightly improved chloasma in 10 of the 12 women (83%, $p < 0.01$) and following 5 months of intake improved or slightly improved chloasma in 6 of the 11 candidates (54%, $p < 0.01$). L* values also increased after GSEP intake (57.8 ± 2.5 at the start vs. 59.3 ± 2.3 at 6 months and 58.7 ± 2.5 at the end of the study). Melanin index significantly decreased after 6 months of the intake (0.025 ± 0.005 at the start vs. 0.019 ± 0.004 at 6 months; $p < 0.01$), and also decreased at the end of study (0.021 ± 0.005; $p < 0.05$) [38]. Another study on the effect of proanthocyanidin for melasma has been reported. Evangeline et al. showed in middle-aged Filipino women the oral intake of cranberry procyanidin and antioxidative vitamin A, C, and E significantly decreased the degree of pigmentation in the left malar and right malar regions ($p < 0.0001$), and the melasma area and severity index showed a significant improvement in the left malar and right malar regions ($p < 0.001$) [39]. The effect of pine bark extract proanthocyanidin (PBE) for human skin has been revealed. Furumura et al. performed a randomized clinical trial of oral supplementation with PBE for middle-aged Japanese women. In the low-dose trial, photoaging scores as assessed by dermatologists after 12 weeks of PBE were compared with those at the beginning of the study. Scores for solar lentigines, mottled pigmentation, roughness, wrinkles, and swelling showed significant improvement at 12 weeks [40].

4. Discussion

We show the scheme for the relationship between the effect of proanthocyanidins and disease field in Figure 2.

Menopausal disorders are caused by estrogen deficiency. Hormone replacement therapy (HRT) remains one of the most effective therapies for vasomotor symptoms that are representative of menopausal symptoms, and it could be beneficial for young and recently postmenopausal women in relation to improvements in cardiovascular health [48]. Soy isoflavones are well known as phytoestrogens and have been heavily reported with regard to the clinical efficacy for menopausal symptoms [49,50]. In this review, we have reported that two types of proanthocyanidins, derived from grape seed and pine bark, improved menopausal symptoms. Terauchi et al. proposed that the mechanism of alleviative effect for vasomotor symptoms is generally attributed to their antioxidant activities, although they do not bind to estrogen receptors, and also have considered that their hypnosedative and anxiolytic activities might partly explain the effects of proanthocyanidins acting through gamma amino butyric acid (GABA)A receptors [51]. Moreover, anxiolytic and depressive activities of proanthocyanidins have been reported, involving the central monoaminergic neurotransmitter systems [52] and inhibiting the expressions of the proinflammatory cytokines, iNOS and COX-2 in the hippocampus [53]. In a prospective cohort study, it has also been revealed that higher proanthocyanidin intakes were associated with lower depression risk, with the highest intake group taking up to 160–179 mg of proanthocyanidins per day. The amount was approximated to the result of intervention trials as mentioned above.

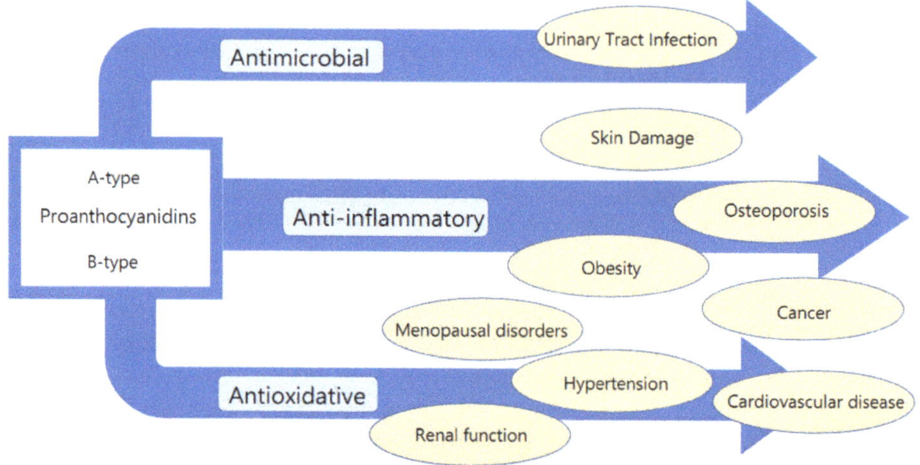

Figure 2. Relationship between the effect of proanthocyanidins and disease field.

Regarding cancer prevention, laboratory experiments using animal models or cultured human cell lines support a potential role of polyphenols in cancer prevention through antioxidant, immunomodulatory, anti-inflammatory, anti-angiogenic, and pro-apoptotic properties [54,55]. However, the effect of polyphenol intake on disease prevention in humans is difficult to predict, partly because in vivo studies often employed doses or concentrations far beyond those achievable by human diet. The prevention of breast and endometrial cancer is of particularly high significance for elderly women. In two cohort studies, it was revealed that the intake of proanthocyanidins was associated with significantly decreased breast and endometrial cancer risk. Proanthocyanidins have antioxidant and antiangiogenesis effects and may influence signal transduction and inhibit the action of DNA topoisomerases [56,57]. Although the bioavailability of higher molecular weight proanthocyanidins is lower, they are characterized by a higher gastric stability [58] and a higher potential scavenger activity [59]. In fact, bioavailability of proanthocyanidins (in monomeric, oligomeric, and polymeric forms of flavan-3-ols) is influenced by their degree of polymerisation; monomers are readily absorbed in the small intestine, whereas oligomers and polymers need to be biotransformed by the colonic microbiota because they are resistant to acid hydrolysis in the stomach [58]. Therefore, phenolic metabolites, rather than the original high-molecular weight compounds found in foods, may be responsible for the health effects derived from proanthocyanidin consumption [60], especially those with higher degree of polymerization. In experimental studies, the microbial metabolites of proanthocyanidins still bearing a free phenolic acid showed protective effects against oxidative stress and obesity [61,62], the major risk factors for endometrial cancer. Culter et al. reported that lung cancer incidence was inversely associated with intakes of proanthocyanidins. In another animal experiment, it was revealed that administration of grape seed proanthocyanidins to athymic nude mice by oral gavage (5 days per week) markedly inhibited the growth of s.c. A549 and H1299 lung tumor xenografts, which was associated with the induction of apoptotic cell death, increased expression of Bax, reduced expression of anti-apoptotic proteins, and activation of caspase-3 in tumor xenograft cells [23]. These results suggest that proanthicyanidins may represent a potential component for lung cancer.

Evidence from many short-term trials in humans has suggested that flavonoids and, in particular, flavanol monomers and procyanidin may have a beneficial effect on blood pressure in humans [63]. In this review, we show the effect of grape seed proanthocyanidin on blood pressure in elderly Japanese women and the relation between proanthocyanidin intake and incidence of hypertension in a large prospective cohort of women. Odai et al. considered that grape seed proanthocyanidin improving blood pressure without affecting flow-mediated dilation indicates that the antioxidant effects of proanthocyanidins could regulate vascular tone, not through NO release, but by other endothelial responses, which results in blood pressure reduction; one study on hypertensive rats that supports their results showed the positive association between reactive oxygen species (ROS) level and pulse wave velocity, arterial wall thickness, and collagen deposition and the beneficial effects of antioxidants on arterial stiffness and remodeling [64], implying that the antioxidant capacity of grape seed proanthocyanidin could contribute to decreased ROS levels and improved vascular elasticity. Evidence for the potential mechanism by which proanthocyanidin would affect blood pressure could also be related to the inhibition of the angiotensin-converting enzyme [65].

Epidemiologic data suggest that dietary flavonoids may have beneficial cardiovascular effects in human populations. Several prospective studies have reported statistically significant inverse associations between total flavonoid intake or the intake of specific classes of flavonoids and cardiovascular disease (CVD) incidence or mortality [66,67]. In this review, we revealed that incremental elastic modulus (Einc) and pulse wave velocity (PWV) also significantly improved by oral intake of grape seed proanthocyanidins, and consumption apple proanthocyanidins produced a statistically significant decrease in oxidized low-density lipoprotein/beta2-glycoprotein I complex (oxLDL-b2GPI) related to atherosclerosis. These results support the fact that proanthocyanidins affect cardiovascular health. In cohort studies, the intake of proanthocyanidins has also been reported to significantly reduce CVD mortality for middle-aged and elderly women. The cardioprotective effects of proanthocyanidins have been highlighted by several studies regarding the mechanism: oxidative stress, cardiomyocytes and the endothelium, anti-inflammatory effects, metabolic effects, etc. [68]. These beneficial effects appear to be mediated by various signaling pathways and mechanisms acting either independently or synergistically.

Overweight and obesity are recognized as important risk factors for type 2 diabetes (T2D). Some cohort studies for elderly women have been investigated with regards to diabetes mellitus and obesity and have revealed that the intake of proanthocyanidins is inversely associated with abdominal obesity or overweight. Yang et al. proposed that proanthocyanidins lower hepatic glucose production by activating adenosine monophosphate(AMP)-activated protein kinase and/or insulin-signaling pathways, play a role in protecting pancreatic β cells from oxidative stress, and promote insulin secretion and β-cell survival, and the actions of proanthocyanidins on liver and pancreatic β cells could lower blood glucose and reduce metabolic and oxidative stress on the islets to enhance glucose homeostasis, which could be of long-term benefit to overall metabolic health [69]. Regarding the hypertension, cardiovascular, and obesity fields, the highest intake group consumed 168–524 mg of proanthocyanidins as a daily average amount in each cohort study reported in this review. These results may suggest that the effective intake of proanthocyanidins for these diseases may be larger than that for menopausal symptoms.

Osteopenia is an important predictor of osteoporosis as it is characterized by low bone mineral density (BMD) [70]. Because osteoporosis has also become an important factor in morbidity and mortality in elderly women, its prevention is therefore of utmost importance in this age group. Many studies have examined the associations between flavonoids and bone health [71–73]; however, most of them, including randomized controlled trials (RCTs), have focused primarily on the isoflavone subclass, which is mainly contained in soy foods. In this review, pine bark proanthocyanidin, not estrogenic compound, supplementation in postmenopausal osteopenic women has been shown to produce favorable effects on bone markers. Moreover, in a cohort study, dietary proanthocyanidin intake was positively associated with BMD in middle-aged and elderly women. The inhibition of receptor activator of nuclear factor-kappa B (NF-κB) ligand (RANKL)-dependent osteoclast differentiation caused by proanthocyanidins was indicated by studies in vitro [74]. Animal experimental results also suggested that proanthocyanidins can promote bone formation [75]. Proanthocyanidins may be expected for the prevention of osteoporosis instead of soy isoflavones.

A urinary tract infection (UTI) is common and increasingly difficult to treat because of the rising rates of antibiotic resistance [76,77]. Approximately 60% of women will experience up to one UTI in their lifetimes. Cranberry consumption has been evaluated as a strategy for reducing clinical UTI recurrence in women with a recent history of a UTI [78,79]. In this review, we have shown that the consumption of cranberry juice or cranberry powder containing proanthocyanidins lowered the incidence of UTI for elderly women. The proanthocyanidins in cranberry have been reported to inhibit the growth of several pathogenic bacteria, such as uropathogenic *Escherichia coli*, cariogenic *Streptococcus mutans*, and oxacillin-resistant *Staphylococcus aureus* [80]. The cranberry proanthocyanidins, consisting primarily of epicatechin tetramers and pentamers with at least one A-type linkage, have been found to be active against the pathogenic bacteria. In addition, the daily intake of 65% cranberry juice is recommended to prevent recurrent cystitis for postmenopausal women in the Guideline in The Japanese Association for Infectious Disease/Japanese Society of Chemotherapy (JAID/JSC) Guidelines for Infection Treatment—Urinary Tract Infection.

Chronic kidney disease (CKD) represents a growing public health issue [81]. Oxidative stress, atherogenesis, nitric oxide homeostasis, and endothelial function play important roles in the pathogenesis of this disease [82–86]. Ageing is associated with structural and functional changes in the kidneys [87], resulting in impaired renal function [88]. Cystatin C provides early indications of renal dysfunction [89]. Ivey et al. indicated in a cohort study that higher intake of proanthocyanidins lowered the plasma cystatin C level. There is direct evidence that proanthocyanidins can specifically improve renal health in animal models by reducing oxidative stress, improving antioxidant defense potential, and reducing oxidative renal injury [56,90,91]. Proanthocyanidins may contribute to the prevention of severe CKD.

Abnormal facial pigmentation such as chloasma (melasma) is often of great cosmetic importance to women. Chloasma is a common acquired symmetrical hypermelanosis characterized by irregular light to dark brown macules and patches on sun-exposed areas of the skin. Although the etiology is unknown, several etiogenic factors have been implicated, including genetic factors, UV exposure, pregnancy, hormonal therapies, cosmetics, phototoxic drugs, and antiseizure medications [92]. In this review, we have reported that grape seed proanthocyanidins and cranberry proanthocyanidin with vitamin A, C, and E improved chloasma in an intervention study. Yamakoshi et al. considered that grape seed proanthocyanidins were likely to inhibit melanogenesis or even melanocyte proliferation only in the chloasma area. The human skin is constantly exposed to UV radiation. UV radiation generates reactive oxygen species (ROS) and leads to oxidative stress. This causes a cascade of erythema and inflammatory reactions, which may be considered as crucial factors affecting the pathogenesis of melasma. Proanthocyanidin has been proven to have a significant free radical scavenging activity in vitro and anti-edema effects in vivo [93]. Additionally, the oral intake of pine bark proanthocyanidins have led to significant reduction in the pigmentation of age spots in photoaged facial skin. The results may be attributed to the antioxidative and anti-inflammatory effects of proanthocyanidins.

The physical, chemical, and biological features of proanthocyanidins depend largely on their structure including the type of flavan-3-ol, particularly on their degree of polymerization. We failed to evaluate the structure–activity relationship (SAR) in each effect because of not so many results of the intervention trial. Proanthocyanidins functioning in terms of absorption and metabolism in human is not yet fully understood. The intestinal cell wall is permeable to proanthocyanidin dimers and trimers, as shown in both in vitro [94] and in vivo experiments [95]. Proanthocyanidins with a degree of polymerisation (DP) below 3 are depolymerised into mixtures of epicatechin monomers and dimers in the acidic environment of the stomach [94] and absorbed by the small intestine [94,95]. However, it is also proposed that a food bolus has a buffering effect, making the acidic conditions milder than that required for proanthocyanidin breakdown. Although proanthocyanidin dimers B1 and B2 can be detected in human plasma [95,96], their absorption is minor, estimated to be more than 100-fold lower than that of the corresponding flavan-3-ol monomers [95]. The cell layer in the intestines is also permeable to oligomeric proanthocyanidins but not polymeric ones [94,95]. Polymeric proanthocyanidins with a DP up to 10 move to the small intestines intact and are mainly degraded by colonic microflora in the cecum and large intestine [97]. Further research is needed to better investigate the active metabolites of proanthocyanidins and detail the mechanism in humans.

5. Conclusions

Proanthocyanidins have many effects such as antioxidant, anti-inflammatory, and antimicrobial activities, with a great diversity of complementary alternative therapy supported by sufficient scientific evidence. The consumption of proanthocyanidins can greatly contribute to health promotion in middle-aged and elderly women.

Funding: This research received no external funding.

Conflicts of Interest: M.T. received a research grants from Ibaraki Prefecture, and personal fees for speaker's bureau from Fuji Pharma Co. Ltd.; Bayer Holding Ltd.; and Hisamitsu Pharmaceutical Co., Inc. The other author has no conflicts of interest to disclose.

References

1. Nelson, H.D. Menopause. *Lancet* **2008**, *371*, 760–770. [CrossRef]
2. Thom, T.; Haase, N.; Rosamond, W. Heart disease and stroke statistics-2006 update: A report from the American Heart Association Statistics Committee and Stroke Statistics Subcommittee. *Circulation* **2006**, *113*, e85–e151.
3. Lobo, R.A. Metabolic syndrome after menopause and the role of hormones. *Maturitas* **2008**, *60*, 10–18. [CrossRef]
4. Derby, C.A.; Crawford, S.; Pasternak, R.C.; Sowers, M.; Sternfeld, B.; Matthews, K.A. Lipid changes during the menopause transition in relation to age and weight: The Study of Women's Health Across the Nation. *Am. J. Epidemiol.* **2009**, *169*, 1352–1361. [CrossRef]
5. Taddei, S. Blood pressure through aging and menopause. *Climacteric* **2009**, *12* (Suppl. S1), 36–40. [CrossRef]
6. Brand, J.S.; van der Schouw, Y.T.; Onland-Moret, N.C. Age at menopause, reproductive life span, and type 2 diabetes risk: Results from the EPIC-InterAct study. *Diabetes Care* **2013**, *36*, 1012–1019. [CrossRef]
7. Ravishankar, D.; Rajora, A.K.; Greco, F.; Osborn, H.M. Flavonoids as prospective compounds for anti-cancer therapy. *Int. J. Biochem. Cell Biol.* **2013**, *45*, 2821–2831. [CrossRef]
8. De la Iglesia, R.; Milagro, F.I.; Campión, J.; Boqué, N.; Martínez, J.A. Healthy properties of proanthocyanidins. *Biofactors* **2010**, *36*, 159–168. [CrossRef]
9. Raufa, A.; Imranb, M.; Abu-Izneidc, T.; Ul-Haqd, I.; Patele, S.; Panf, X.; Nazg, S.; Silvah, A.S.; Saeedi, F.; Suleriaj, H.A.R. Proanthocyanidins: A comprehensive review. *Biomed. Pharmacother.* **2019**, *116*, 108999. [CrossRef]
10. Hellstorm, J.K.; Torronen, A.R.; Mattila, P.H. Proanthocyanidins in Common Food Products of Plant Origin. *J. Agric. Food Chem.* **2009**, *57*, 7899–7906. [CrossRef]
11. Gu, L.; Kelm, M.A.; Hammerstone, J.F.; Beecher, G.; Holden, J.; Haytowitz, D.; Prior, R.L. Screening of foods containing proanthocyanidins and their characterization using LC-MS/MS and thiolytic degradation. *J. Agric. Food Chem.* **2003**, *51*, 7513–7521. [CrossRef]
12. Gu, L.; Kelm, M.A.; Hammerstone, J.F.; Beecher, G.; Holden, J.; Haytowitz, D.; Gebhardt, S.; Prior, R.L. Concentrations of proanthocyanidins in common foods and estimations of normal consumption. *J. Nutr.* **2004**, *134*, 613–617. [CrossRef]
13. Smeriglio, A.; Barreca, D.; Bellocco, E.; Trombetta, D. Proanthocyanidins and hydrolysable tannins: Occurrence, dietary intake and pharmacological Effects. *Br. J. Pharmacol.* **2017**, *174*, 1244–1262. [CrossRef]
14. Yang, L.; Xian, D.; Xiong, X.; Lai, R.; Song, J.; Zhong, J. Proanthocyanidins against Oxidative Stress: From Molecular Mechanisms to Clinical Applications. *Biomed. Res. Int.* **2018**, *2018*, 8584136. [CrossRef]

15. González-Quilen, C.; Rodríguez-Gallego, E.; Beltrán-Debón, R.; Pinent, M.; Ardévol, A.; Blay, M.T.; Terra, X. Health-Promoting Properties of Proanthocyanidins for Intestinal Dysfunction. *Nutrients* **2020**, *12*, 130. [CrossRef]

16. Ferraz da Costa, D.C.; Pereira Rangel, L.; Quarti, J.; Santos, R.A.; Silva, J.L.; Fialho, E. Bioactive Compounds and Metabolites from Grapes and Red Wine in Breast Cancer Chemoprevention and Therapy. *Molecules* **2020**, *25*, 3531. [CrossRef]

17. Terauchi, M.; Horiguchi, N.; Kajiyama, A.; Akiyoshi, M.; Owa, Y.; Kato, K.; Kubota, T. Effects of grape seed proanthocyanidin extract on menopausal symptoms, body composition, and cardiovascular parameters in middle-aged women: A randomized, double-blind, placebo-controlled pilot study. *Menopause* **2014**, *21*, 990–996. [CrossRef]

18. Kohama, T.; Negami, M. Effect of Low-dose French Maritime Pine Bark Extract on Climacteric Syndrome in 170 Perimenopausal Women A Randomized, Double-blind, Placebo-controlled Trial. *J. Reprod. Med.* **2013**, *58*, 39–46.

19. Yang, H.M.; Liao, M.F.; Zhu, S.Y.; Liao, M.N.; Rohdewald, P. A randomised, double-blind, placebo-controlled trial on the effect of Pycnogenol® on the climacteric syndrome in peri-menopausal women. *Acta Obstet. Gynecol.* **2007**, *86*, 978–985. [CrossRef]

20. Chang, S.C.; Cassidy, A.; Willett, W.C.; Rimm, E.B.; O'Reilly, E.J.; Okereke, O.I. Dietary flavonoid intake and risk of incident depression in midlife and older women. *Am. J. Clin. Nutr.* **2016**, *104*, 704–714. [CrossRef]

21. Touvier, M.; Druesne-Pecollo, N.; Kesse-Guyot, E.; Andreeva, V.A.; Fezeu, L.; Galan, P.; Serge Hercberg, S.; Latino-Martel, P. Dual association between polyphenol intake and breast cancer risk according to alcohol consumption level: A prospective cohort study. *Breast Cancer Res. Treat.* **2013**, *137*, 225–236. [CrossRef]

22. Rossi, M.; Edefonti, V.; Parpinel, M.; Lagiou, P.; Franchi, M.; Ferraroni, M.; Decarli, A.; Zucchetto, A.; Serraino, D.; Maso, L.D.; et al. Proanthocyanidins and other flavonoids in relation n to endometrial cancer risk: A case–control study in Italy. *Br. J. Cancer* **2013**, *109*, 1914–1920. [CrossRef]

23. Cutler, G.J.; Jennifer, A.; Nettleton, J.A.; Ross, J.A.; Harnack, L.J.; Jacobs, D.R., Jr.; Scrafford, C.G.; Barraj, L.M.; Pamela, J.; Mink, P.J.; et al. Dietary flavonoid intake and risk of cancer in postmenopausal women: The Iowa Women's Health Study. *Int. J. Cancer* **2008**, *123*, 664–671. [CrossRef]

24. Odai, T.; Terauchi, M.; Kato, K.; Asuka Hirose, A.; Miyasaka, N. Effects of Grape Seed Proanthocyanidin Extract on Vascular Endothelial Function in Participants with Prehypertension: A Randomized, Double-Blind, Placebo-Controlled Study. *Nutrients* **2019**, *11*, 2844. [CrossRef]

25. Lajous, M.; Rossignol, E.; Fagherazzi, G.; Perquier, F.; Scalbert, A.; Clavel-Chapelon, F.; Boutron-Ruault, M.C. Flavonoid intake and incident hypertension in women. *Am. J. Clin. Nutr.* **2016**, *103*, 1091–1098. [CrossRef]

26. Zhao, S.; Bomser, J.; Joseph, E.L.; DiSilvestro, R.A. Intakes of apples or apple polyphenols decease plasma values for oxidized low-density lipoprotein/beta2-glycoprotein I complex. *J. Funct. Food* **2013**, *5*, 493–497. [CrossRef]

27. Mink, P.J.; Scrafford, C.G.; Barraj, L.M.; Harnack, L.; Hong, C.-P.; Nettleton, J.A.; Jacobs, D.R., Jr. Flavonoid intake and cardiovascular disease mortality: A prospective study in postmenopausal women. *Am. J. Clin. Nutr.* **2007**, *85*, 895–909. [CrossRef]

28. McCullough, M.L.; Peterson, J.J.; Patel, R.; Jacques, P.F.; Shah, R.; Dwyer, J.T. Flavonoid intake and cardiovascular disease mortality in a prospective cohort of US adults. *Am. J. Clin. Nutr.* **2012**, *95*, 454–464. [CrossRef]

29. Jennings, A.; MacGregor, A.; Spector, T.; Cassidy, A. Higher dietary flavonoid intakes are associated with lower objectively measured body composition in women: Evidence from discordant monozygotic twins. *Am. J. Clin. Nutr.* **2017**, *105*, 626–634. [CrossRef]

30. Tresserra-Rimbau, A.; Castro-Barquero, S.; Vitelli-Storelli, F.; Becerra-Tomas, N.; Vázquez-Ruiz, Z.; Díaz-López, A.; Corella, D.; Castañer, O.; Romaguera, D.; Jesús Vioque, J.; et al. Associations between Dietary Polyphenols and Type 2 Diabetes in a Cross-Sectional Analysis of the PREDIMED-Plus Trial: Role of Body Mass Index and Sex. *Antioxidants* **2019**, *8*, 537. [CrossRef]

31. Kim, S.-A.; Kim, J.; Jun, S.; Wie, G.-A.; Shin, S.; Joung, H. Association between dietary flavonoid intake and obesity among adults in Korea. *Appl. Physiol. Nutr. Metab.* **2020**, *45*, 203–212. [CrossRef]

32. Panahande, S.B.; Maghbooli, Z.; Hossein-nezhad4, A.; Qorbani, M.; Moeini-Nodeh, S.; Haghi-Aminjan, H.; Hosseini, S. Effects of French maritime pine bark extract (Oligopin®) supplementation on bone remodeling markers in postmenopausal osteopenic women: A randomized clinical trial. *Phytother. Res.* **2019**, *33*, 1233–1240. [CrossRef]

33. Zhang, Z.-Q.; He, L.-P.; Liu, Y.-H.; Liu, J.; Su, Y.-X.; Chen, Y.-M. Association between dietary intake of flavonoid and bone mineral density in middle aged and elderly Chinese women and men. *Osteoporos. Int.* **2014**, *25*, 2417–2425. [CrossRef]

34. Takahashi, S.; Hamasuna, R.; Yasuda, M.; Arakawa, S.; Tanaka, K.; Ishikawa, K.; Hiroshi Kiyota, H.; Hayami, H.; Yamamoto, S.; Kubo, T.; et al. A randomized clinical trial to evaluate the preventive effect of cranberry juice (UR65) for patients with recurrent urinary tract infection. *J. Infect. Chemother.* **2013**, *19*, 112–117. [CrossRef]

35. Maki, K.C.; Kaspar, K.L.; Khoo, C.; Derrig, L.H.; Schild, A.L.; Gupta, K. Consumption of a cranberry juice beverage lowered the number of clinical urinary tract infection episodes in women with a recent history of urinary tract infection. *Am. J. Clin. Nutr.* **2016**, *103*, 1434–1442. [CrossRef]

36. Vostalova, J.; Vidlar, A.; Simanek, V.; Galandakova, A.; Kosina, P.; Vacek, J.; Jana Vrbkova, J.; Zimmermann, B.F.; Ulrichova, J.; Student, V. Are High Proanthocyanidins Key to Cranberry Efficacy in the Prevention of Recurrent Urinary Tract Infection? *Phytother. Res.* **2015**, *29*, 1559–1567. [CrossRef]

37. Ivey, K.L.; Lewis, J.R.; Lim, W.H.; Lim, E.M.; Hodgson, J.M.; Prince, R.L. Associations of Proanthocyanidin Intake with Renal Function and Clinical Outcomes in Elderly Women. *PLoS ONE* **2013**, *8*, e71166. [CrossRef]

38. Yamakoshi, J.; Sano, A.; Tokutake, S.; Saito, M.; Kikuchi, M.; Kubota, Y.; Kawachi, Y.; Otsuka, F. Oral Intake of Proanthocyanidin-Rich Extract from Grape Seeds Improves Chloasma. *Phytother. Res.* **2004**, *18*, 895–899. [CrossRef]

39. Handog, E.B.; Galang, D.A.V.F.; Leon-Godinez, M.A.; Chan, G.P. A randomized, double-blind, placebo-controlled trial of oral procyanidin with vitamins A, C, E for melasma among Filipino women. *Int. J. Dermat.* **2009**, *48*, 896–901. [CrossRef]

40. Furumura, M.; Sato, N.; Kusaba, N.; Takagaki, K.; Nakayama, J. Oral administration of French maritime pine bark extract (Flavangenol®) improves clinical symptoms in photoaged facial skin. *Clin. Interv. Aging* **2012**, *7*, 275–286. [CrossRef]

41. Grosso, G.; Micek, A.; Godos, J.; Pajak, A.; Sciacca, S.; Galvano, F.; Giovannucci, E.L. Dietary Flavonoid and Lignan Intake and Mortality in Prospective Cohort Studies: Systematic Review and Dose-Response Meta-Analysis. *Am. J. Epidemiol.* **2017**, *185*, 1304–1316. [CrossRef]

42. Goetz, M.E.; Judd, S.E.; Safford, M.M.; Hartman, T.J.; McClellan, W.M. Vaccarino, Dietary flavonoid intake and incident coronary heart disease: The REasons for Geographic and Racial Differences in Stroke (REGARDS) study. *Am. J. Clin. Nutr.* **2016**, *104*, 1236–1244. [CrossRef]

43. Wang, X.; Ouyang, Y.Y.; Liu, J.; Zhao, G. Flavonoid intake and risk of CVD: A systematic review and meta-analysis of prospective cohort studies. *Br. J. Nutr.* **2014**, *111*, 1–11. [CrossRef]

44. Raman, G.; Shams-White, M.; Avendano, E.E.; Chen, F.; Novotny, J.A.; Cassidy, A. Dietary intake of flavan-3-ols and cardiovascular health: A field synopsis using evidence Mapping of randomized trials and prospective cohort studies. *Syst. Rev.* **2018**, *7*, 100. [CrossRef]

45. Raman, G.; Avendano, E.E.; Chen, S.; Wang, J.; Matson, J.; Gayer, B.; Novotny, J.A.; Cassidy, A. Dietary intakes of flavan-3-ols and cardiometabolic health: Systematic review and meta-analysis of randomized trials and prospective cohort studies. *Am. J. Clin. Nutr.* **2019**, *110*, 1067–1078. [CrossRef]

46. Bertoia, M.L.; Rimm, E.B.; Mukamal, K.J.; Hu, F.B.; Willett, W.C.; Cassidy, A. Dietary flavonoid intake and weight maintenance: Three prospective cohorts of 124086 US men and women followed for up to 24 years. *BMJ* **2016**, *352*, i17. [CrossRef]

47. Wedick, N.M.; Pan, A.; Cassidy, A.; Rimm, E.B.; Sampson, L.; Rosner, B.; Willett, W.; Hu, F.B.; Sun, Q.; van Dam, R.M. Dietary flavonoid intakes and risk of type 2 diabetes in US men and women. *Am. J. Clin. Nutr.* **2012**, *95*, 925–933. [CrossRef]

48. Chen, L.-R.; Ko, N.-Y.; Chen, K.-H.; Rossouw, J.E.; Prentice, R.L.; Manson, J.E. Postmenopausal hormone therapy and risk of cardiovascular disease by age and years since menopause. *JAMA* **2007**, *297*, 1465–1477.

49. Chen, L.-R.; Ko, N.-Y.; Chen, K.-H. Isoflavone Supplements for Menopausal Women: A Systematic Review. *Nutrients* **2019**, *11*, 2649. [CrossRef]

50. Hirose, A.; Terauchi, M.; Akiyoshi, M.; Owa, Y.; Kato, K.; Kubota, T. Low-dose isoflavone aglycone alleviates psychological symptoms of menopause in Japanese women: A randomized, double-blind, placebo-controlled study. *Arch. Gynecol. Obstet.* **2016**, *293*, 609–615. [CrossRef]

51. Moreira, E.L.; Rial, D.; Duarte, F.S. Central nervous system activity of the proanthocyanidin-rich fraction obtained from Croton celtidifolius in rats. *J. Pharm. Pharmacol.* **2010**, *62*, 1061–1068. [CrossRef]

52. Xu, Y.; Li, S.; Chen, R.; Li, G.; Barish, P.A.; You, W.; Chen, L.; Lin, M.; Ku, B.; Pan, J.; et al. Antidepressant-like effect of low molecular proanthocyanidin in mice: Involvement of monoaminergic system. *Pharmacol. Biochem. Behav.* **2010**, *94*, 447–453. [CrossRef]

53. Jiangb, X.; Liua, J.; Lin, Q.; Mao, K.; Tian, F.; Jing, C.; Wang, C.; Ding, L.; Pang, C. Proanthocyanidin prevents lipopolysaccharide-induced depressive-like behavior in mice via neuroinflammatory pathway. *Brain Res. Bull.* **2017**, *135*, 40–46. [CrossRef]

54. Lambert, J.D.; Hong, J.; Yang, G.Y.; Liao, J.; Yang, C.S. Inhibition of carcinogenesis by polyphenols: Evidence from laboratory investigations. *Am. J. Clin. Nutr.* **2005**, *81*, 284S–291S. [CrossRef]

55. Yang, C.S.; Wang, X.; Lu, G.; Picinich, S.C. Cancer prevention by tea: Animal studies, molecular mechanisms and human relevance. *Nat. Rev. Cancer* **2009**, *9*, 429–439. [CrossRef]

56. Bagchi, D.; Bagchi, M.; Stohs, S.J.; Das, D.K.; Ray, S.D.; Kuszynski, C.A.; Joshi, S.S.; Pruess, H.G. Free radicals and grape seed proanthocyanidin extract: Importance in human health and disease prevention. *Toxicology* **2000**, *148*, 187–197. [CrossRef]

57. Wang, Y.H.; Ge, B.; Yang, X.L.; Zhai, J.; Yang, L.N.; Wang, X.X.; Liu, X.; Shi, J.C.; Wu, Y.J. Proanthocyanidins from grape seeds modulates the nuclear factor kappa B signal transduction pathways in rats with TNBS-induced recurrent ulcerative colitis. *Int. Immunopharmacol.* **2011**, *11*, 1620–1627. [CrossRef]

58. Krook, M.A.; Hagerman, A.E. Stability of polyphenols epigallocatechin gallate and pentagalloyl glucose in a simulated digestive system. *Food Res. Int.* **2012**, *49*, 112–116. [CrossRef]

59. Hagerman, A.E.; Riedl, K.M.; Jones, G.A.; Sovik, K.N.; Ritchard, N.T.; Hartzfeld, P.W.; Riechel, T.L. High molecular weight plant polyphenolics (tannins) as biological antioxidants. *J. Agric. Food Chem.* **1998**, *46*, 1887–1892. [CrossRef]

60. Monagas, M.; Urpi-Sarda, M.; Sanchez-Patan, F.; Llorach, R.; Garrido, I.; Gomez-Cordoves, C.; Andres-Lacueva, C.; Bartolome, B. Insights into the metabolism and microbial biotransformation of dietary flavan-3-ols and the bioactivity of their metabolites. *Food Funct.* **2010**, *1*, 233–253. [CrossRef]

61. Gonthier, M.P.; Donovan, J.L.; Texier, O.; Felgines, C.; Remesy, C.; Scalbert, A. Metabolism of dietary procyanidins in rats. *Free Radic. Biol. Med.* **2003**, *35*, 837–844. [CrossRef]

62. Thom, E. The effect of chlorogenic acid enriched coffee on glucose absorption in healthy volunteers and its effect on body mass when used long-term in overweight and obese people. *J. Int. Med. Res.* **2007**, *35*, 900–908. [CrossRef]

63. Kay, C.D.; Hooper, L.; Kroon, P.A.; Rimm, E.B.; Cassidy, A. Relative impact of flavonoid composition, dose and structure on vascular function: A systematic review of randomised controlled trials of flavonoid-rich food products. *Mol. Nutr. Food Res.* **2012**, *56*, 1605–1616. [CrossRef]

64. Chen, Q.Z.; Han, W.Q.; Chen, J.; Zhu, D.L.; Chen-Yan Gao, P.J. Anti-stiffness effect of apocynin in deoxycorticosterone acetate-salt hypertensive rats via inhibition of oxidative stress. *Hypertens. Res.* **2013**, *36*, 306–312. [CrossRef]

65. Actis-Goretta, L.; Ottaviani, J.I.; Keen, C.L.; Fraga, C.G. Inhibition of angiotensin converting enzyme (ACE) activity by flavan-3-ols and procyanidins. *FEBS Lett.* **2003**, *555*, 597–600. [CrossRef]

66. Hertog, M.G.; Kromhout, D.; Aravanis, C. Flavonoid intake and long-term risk of coronary heart disease and cancer in the seven countries study. *Arch. Intern. Med.* **1995**, *155*, 381–386. [CrossRef]

67. Yochum, L.; Kushi, L.H.; Meyer, K.; Folsom, A.R. Dietary flavonoid intake and risk of cardiovascular disease in postmenopausal women. *Am. J. Epidemiol.* **1999**, *149*, 943–949. [CrossRef]

68. Kruger, M.J.; Davies, N.; Myburgh, K.H.; Lecour, S. Proanthocyanidins, anthocyanins and cardiovascular diseases. *Food Res. Int.* **2014**, *59*, 41–52. [CrossRef]

69. Yang, K.; Chan, C.B. Proposed mechanisms of the effects of proanthocyanidins on glucose homeostasis. *Nutr. Rev.* **2017**, *75*, 8–642. [CrossRef]

70. Melton, L.J., III; Chrischilles, E.A.; Cooper, C.; Lane, A.W.; Riggs, B.L. How many women have osteoporosis? *J. Bone Miner. Res.* **2005**, *20*, 886–892. [CrossRef]

71. Liu, J.; Ho, S.C.; Su, Y.X.; Chen, W.Q.; Zhang, C.X.; Chen, Y.M. Effect of long-term intervention of soy isoflavones on bone mineral density in women: A meta-analysis of randomized controlled trials. *Bone* **2009**, *44*, 948–953. [CrossRef]

72. Hardcastle, A.C.; Aucott, L.; Reid, D.M.; Macdonald, H.M. Associations between dietary flavonoid intakes and bone health in a Scottish population. *J. Bone Miner. Res.* **2011**, *26*, 941–947. [CrossRef]

73. Welch, A.; MacGregor, A.; Jennings, A.; Fairweather-Tait, S.; Spector, T.; Cassidy, A. Habitual flavonoid intakes are positively associated with bone mineral density in women. *J. Bone Miner. Res.* **2012**, *27*, 1872–1878. [CrossRef]

74. Tanabe, S.; Santos, J.; La, V.D.; Howell, A.B.; Grenier, D. A-type cranberry proanthocyanidins inhibit the RANKL-dependent differentiation and function of human osteoclasts. *Molecules* **2011**, *16*, 2365–2374. [CrossRef]

75. Ishikawa, M.; Maki, K.; Tofani, I.; Kimura, K.; Kimura, M. Grape seed proanthocyanidins extract promotes bone formation in rat's mandibular condyle. *Eur. J. Oral Sci.* **2005**, *113*, 47–52. [CrossRef]

76. Foxman, B.; Barlow, R.; D'Arcy, H.; Gillespie, B.; Sobel, J.D. Urinary tract infection: Self-reported incidence and associated costs. *Ann. Epidemiol.* **2000**, *10*, 509–515. [CrossRef]

77. Litwin, M.S.; Saigal, C.S.; Yano, E.M.; Avila, C.; Gescgwind, S.A.; Hanley, J.M.; Joyce, G.F.; Madison, R.; Pace, J.; Polich, S.M. Urologic Diseases in America Project: Analytical methods and principal findings. *J. Urol.* **2005**, *173*, 933–937. [CrossRef]

78. Wang, C.H.; Fang, C.C.; Chen, N.C.; Liu, S.S.; Yu, P.H.; Wu, T.Y.; Chen, W.T.; Lee, C.C.; Chen, S.C. Cranberry-containing products for prevention of urinary tract infections in susceptible populations: A systemic review and meta-analysis of randomized controlled trials. *Arch. Intern. Med.* **2012**, *172*, 988–996. [CrossRef]

79. Jepson, R.G.; Williams, G.; Craig, J.C. Cranberries for preventing urinary tract infections. *Cochrane Database Syst. Rev.* **2012**, *10*, CD001321. [CrossRef]

80. Côté, J.; Caillet, S.; Doyon, G.; Sylvain, J.F.; Lacroix, M. Bioactive compounds in cranberries and their biological properties. *Crit. Rev. Food Sci. Nutr.* **2010**, *50*, 666–679.

81. Coresh, J.; Selvin, E.; Stevens, L.A.; Manzi, J.; Kusek, J.W. Prevalence of chronic kidney disease in the United States. *J. Am. Med. Assoc.* **2007**, *298*, 2038–2047. [CrossRef] [PubMed]

82. Annuk, M.; Zilmer, M.; Lind, L.; Linde, T.; Fellstro"m, B. Oxidative stress and endothelial function in chronic renal failure. *J. Am. Soc. Nephrol.* **2001**, *12*, 2747–2752. [PubMed]

83. Kielstein, J.T.; Frölich, J.C.; Haller, H.; Fliser, D. ADMA (asymmetric dimethylarginine): An atherosclerotic disease mediating agent in patients with renal disease? *Nephrol. Dial. Transplant.* **2001**, *16*, 1742–1745. [CrossRef] [PubMed]

84. Stenvinkel, P.; Heimburger, O.; Paultre, F.; Diczfalusy, U.; Wang, T. Strong association between malnutrition, inflammation, and atherosclerosis in chronic renal failure. *Kidney Int.* **1999**, *55*, 1899–1911. [CrossRef]

85. Bagi, Z.; Hamar, P.; Antus, B.; Rosivall, L.; Koller, A. Chronic renal failure leads to reduced flow-dependent dilation in isolated rat skeletal muscle arterioles due to lack of NO mediation. *Kidney Blood Press. Res.* **2003**, *26*, 19–26. [CrossRef]

86. Liao, J.; Bettmann, M.; Sandor, T.; Tucker, J.; Coleman, S. Differential impairment of vasodilator responsiveness of peripheral resistance and conduit vessels in humans with atherosclerosis. *Circ. Res.* **1991**, *68*, 1027–1034. [CrossRef]

87. Anderson, S.; Eldadah, B.; Halter, J.B.; Hazzard, W.R.; Himmelfarb, J. Acute kidney injury in older adults. *J. Am. Soc. Nephrol.* **2011**, *22*, 28–38. [CrossRef]

88. Silva, F.G. The aging kidney: A review—Part I. *Int. Urol. Nephrol.* **2005**, *37*, 185–205. [CrossRef]

89. Bökenkamp, A.; Herget-Rosenthal, S.; Bökenkamp, R. Cystatin C, kidney function and cardiovascular disease. *Pediatr. Nephrol.* **2006**, *21*, 1223–1230. [CrossRef]

90. Yanarates, O.; Guven, A.; Sizlan, A.; Uysal, B.; Akgul, O. Ameliorative effects of proanthocyanidin on renal ischemia/reperfusion injury. *Ren. Fail.* **2008**, *30*, 931–938. [CrossRef]

91. Lee, Y.A.; Kim, Y.J.; Cho, E.J.; Yokozawa, T. Ameliorative effects of proanthocyanidin on oxidative stress and inflammation in streptozotocin induced diabetic rats. *J. Agric. Food Chem.* **2007**, *55*, 9395–9400. [CrossRef] [PubMed]

92. Kang, W.H.; Yoon, K.H.; Lee, E.-S. Melasma: Histopathological characteristics in 56 Korean patients. *Br. J. Dermatol.* **2002**, *146*, 228–237. [CrossRef] [PubMed]

93. Blazsó, G.; Gabor, M.; Rohdewald, P. Antiinflammatory activities and superoxide radical scavenging activities of a procyanidin containing extract from the bark of Pinus pinaster Sol. and its fractions. *Pharm. Pharmacol. Lett.* **1994**, *3*, 217–220.

94. Deprez, S.; Mila, I.; Huneau, J.F.; Tome, D.; Scalbert, A. Transport of Proanthocyanidin dimer, trimer, and polymer across monolayers of human intestinal epithelial Caco-2 cells. *Antioxid. Redox Signal.* **2001**, *3*, 957–967. [CrossRef] [PubMed]

95. Holt, R.R.; Lazarus, S.A.; Sullards, M.C.; Zhu, Q.Y.; Schramm, D.D.; Hammerstone, J.F. Procyanidin dimer B2 (epicatechin-(4beta-8)-epicatechin) in human plasma after the consumption of a flavanol-rich cocoa. *Am. J. Clin. Nutr.* **2002**, *76*, 798–804. [CrossRef] [PubMed]

96. Sano, A.; Yamakoshi, J.; Tokutake, S.; Tobe, K.; Kubota, Y.; Kikuchi, M. Procyanidin B1 is detected in human serum after intake of proanthocyanidin-rich grape seed extract. *Biosci. Biotechnol. Biochem.* **2003**, *67*, 1140–1143. [CrossRef] [PubMed]

97. Rios, L.Y.; Bennett, R.N.; Lazarus, S.A.; Remesy, C.; Scalbert, A.; Williamson, G. Cocoa procyanidins are stable during gastric transit in humans. *Am. J. Clin. Nutr.* **2002**, *76*, 1106–1110. [CrossRef]

Publisher's Note: MDPI stays neutral with regard to jurisdictional claims in published maps and institutional affiliations.

 © 2020 by the authors. Licensee MDPI, Basel, Switzerland. This article is an open access article distributed under the terms and conditions of the Creative Commons Attribution (CC BY) license (http://creativecommons.org/licenses/by/4.0/).

Review

Evaluation of Clinical Meaningfulness of Red Clover (*Trifolium pratense* L.) Extract to Relieve Hot Flushes and Menopausal Symptoms in Peri- and Post-Menopausal Women: A Systematic Review and Meta-Analysis of Randomized Controlled Trials

Wiesław Kanadys [1], Agnieszka Barańska [1,*], Agata Błaszczuk [2], Małgorzata Polz-Dacewicz [2], Bartłomiej Drop [1], Krzysztof Kanecki [3] and Maria Malm [1]

[1] Department of Informatics and Medical Statistics, Medical University of Lublin, 20-090 Lublin, Poland; wieslaw.kanadys@wp.pl (W.K.); bartlomiej.drop@umlub.pl (B.D.); maria.malm@umlub.pl (M.M.)

[2] Department of Virology with SARS Laboratory, Medical University of Lublin, 20-093 Lublin, Poland; agata.blaszczuk@umlub.pl (A.B.); malgorzata.polz.dacewicz@umlub.pl (M.P.-D.)

[3] Department of Social Medicine and Public Health, Warsaw Medical University, 02-007 Warsaw, Poland; kkanecki@wum.edu.pl

* Correspondence: agnieszkabaranska@umlub.pl

Citation: Kanadys, W.; Barańska, A.; Błaszczuk, A.; Polz-Dacewicz, M.; Drop, B.; Kanecki, K.; Malm, M. Evaluation of Clinical Meaningfulness of Red Clover (*Trifolium pratense* L.) Extract to Relieve Hot Flushes and Menopausal Symptoms in Peri- and Post-Menopausal Women: A Systematic Review and Meta-Analysis of Randomized Controlled Trials. *Nutrients* **2021**, *13*, 1258. https://doi.org/10.3390/nu13041258

Academic Editor: Masakazu Terauchi

Received: 19 March 2021
Accepted: 9 April 2021
Published: 11 April 2021

Publisher's Note: MDPI stays neutral with regard to jurisdictional claims in published maps and institutional affiliations.

Copyright: © 2021 by the authors. Licensee MDPI, Basel, Switzerland. This article is an open access article distributed under the terms and conditions of the Creative Commons Attribution (CC BY) license (https://creativecommons.org/licenses/by/4.0/).

Abstract: The meta-analysis presented in this article covered the efficacy of red clover isoflavones in relieving hot flushes and menopausal symptoms in perimenopausal and postmenopausal women. Studies were identified by MEDLINE (PubMed), Embase, and the Cochrane Library searches. The quality of the studies was evaluated according to Cochrane criteria. A meta-analysis of eight trials (ten comparisons) demonstrated a statistically significant reduction in the daily incidence of hot flushes in women receiving red clover compared to those receiving placebo: weighted mean difference (WMD—weighted mean difference) -1.73 hot flushes per day, 95% CI (confidence interval) -3.28 to -0.18; $p = 0.0292$. Due to 87.34% heterogeneity, the performed analysis showed substantive difference in comparisons of postmenopausal women with ≥ 5 hot flushes per day, when the follow-up period was 12 weeks, with an isoflavone dose of ≥ 80 mg/day, and when the formulations contained a higher proportion of biochanin A. The meta-analysis of included studies assessing the effect of red clover isoflavone extract on menopausal symptoms showed a statistically moderate relationship with the reduction in the daily frequency of hot flushes. However, further well-designed studies are required to confirm the present findings and to finally determine the effects of red clover on the relief of flushing episodes.

Keywords: red clover; isoflavones; Trifolium pratense; hot flushes; menopausal symptoms; postmenopausal women; perimenopausal women

1. Introduction

Menopause is characterized by amenorrhea due to the cessation of ovarian function. The decrease in circulating estrogens levels can induce menopausal disorders, including shorter-term symptoms, such as vasomotor symptoms, palpitations, sleep difficulties, headaches, fatigue, mood disturbances, and impaired concentration, and longer-term chronic conditions, such as cardiovascular diseases, accelerated bone loss, and cognitive impairment [1,2].

Hot flushes (HFs) are the most common symptom of menopause in about 70% of women, with differences in different populations, and may persist for several years after menopause [3,4]. The frequency of these symptoms depends on several factors, including climate, race/ethnicity, diet, lifestyle, women's roles, and attitudes regarding the end of reproductive life and aging [5–7]. They can affect not only the quality of life, but also contribute to sleep and mood disturbance, which can potentially affect daily activities at home and at work to such an extent that treatment is required [8,9]. HFs are thought to be the result of the brain's response to a progressive estrogen deficiency and fluctuation in the activity of neurotransmitters, especially in the serotonergic and noradrenergic pathways, which leads to instability of the mechanism of thermoregulation in the hypothalamus. Ultimately, this leads to increased blood flow through the skin and increased sweat gland activity, and as a result, these symptoms appear [10,11].

Despite the well-known benefits of menopausal hormonal therapy (HT), due to potentially serious side effects and breast cancer risk, the use of this therapy, even in the treatment of HFs, remains controversial [12,13]. Many women discontinued HT after the publication of the results of the Women's Health Initiative, looking for an effective and safe alternative to relieve menopausal symptoms [14,15]. The lack of acceptance of HT, related to concerns about its safety, has led to the popularization of many alternative and complementary methods of treatment [16–18]. For some years, red clover has been one such alternative used by women to treat vasomotor symptoms of menopause [19].

Red clover (*Trifolium pratense L., Fabaceae*) mainly contains the isoflavone aglycones, formononetin, and biochanin A; other isoflavones, such as genistein, daidzein, glycitein, and prunetin, were also identified in small quantities [19,20]. The mentioned methoxy precursors in the intestine and liver are demethylated by cytochrome P450 isozymes to the active forms, genistein and daidzein [21]. Red clover isoflavones with structural similarities to the endogenous 17β-estradiol reveal their biological effects via activating estrogen receptors (ER), with a higher affinity to ER-β in comparison to ER-α. In addition, a number of non-hormonal effects have been reported in isoflavones, including tyrosine kinase inhibition, antioxidant activity, and effects on ion transport [18,22,23]. In addition, a number of non-hormonal effects have been reported in isoflavones, including antioxidant activity, tyrosine kinase inhibition, and effects on ion transport [18,22,23].

This systematic review with meta-analysis aimed to clarify whether supplementation of red clover isoflavone extract (RCIE) affects menopausal symptoms in perimenopausal and postmenopausal women.

2. Materials and Methods

2.1. Search Strategy and Study Selection

To determine if intervention with the RCIE, as compared to placebo, relieves HFs, we reviewed published clinical trials in accordance with the PRISMA (The Preferred Reporting Items for Systematic Reviews and Meta-Analyses) guidelines [24]. The electronic databases MEDLINE (PubMed), Embase, and the Cochrane Library were searched for the identification of randomized controlled trials from 1999 to January 2020. The following search terms were used for all databases in various combination: "menopause" or "perimenopause" or "postmenopause" AND "red clover" or "Trifolium pratense" or "isoflavone" AND "hot flushes" or "menopausal symptoms". The search was limited to the articles published in the English language for full analysis. References to selected research and review articles related to the topic of the work were also searched in order to identify additional studies.

2.2. Inclusion and Exclusion Criteria

Studies were considered eligible for inclusion if they met all of the following criteria: (a) parallel-group controlled trials; (b) trials with a crossover design that contained data for the first period; (c) comparison with placebo; (d) perimenopausal and postmenopausal women; (e) experiencing moderate to severe HFs at least three times per day in a two-week follow-up before the study entry; (f) primary outcomes that were changes in frequency of HFs per day, obtained by self-report using symptom diaries; (g) secondary outcomes that were the cumulative rating of menopausal symptoms using a questionnaire based on the respondents' replies concerning the intensity of complaints. The used questionnaires and their descriptions are as follows: the Kupperman Menopausal Index (KMI) is a measure using a list of 11 symptoms (hot flushes, excessive sweating, sleep disturbances, irritability, depressive mood, attention deficit disorder, joint and bone pain, headache, arrhythmias, paresthesia) rated on a 4-point severity scale [25]. The Greene Climacteric Scale (GCS) is a tool based on a list of 21 symptoms rated on a 4-point Likert scale. The symptoms are divided into three categories: (a) psychological—anxiety (heart beating fast or strong, feeling tense or nervous, difficulty sleeping, being agitated, having anxiety or panic attacks, difficulty concentrating) and depression (feeling tired or lacking energy, loss of interest in most things, feeling unhappy or depressed, crying, irritable); (b) somatic (dizziness or fainting, pressure or tightness in the head, numbness in part of the body, headaches, aches and pains in the muscles and joints, loss of feeling in the hands or feet, difficulty breathing); (c) vasomotor (hot flushes, sweating at night); with an additional question related to sexual dysfunction (loss of interest in sex) [26]. The Menopause Rating Scale (MRS) consists of 11 items divided into three subscales, i.e., sweating/hot flushes, heart discomfort, sleep problems, joint and muscle problems, classified as somatic-vegetative symptoms; depressed mood, irritability, restlessness, and physical/mental exhaustion, classified as psychological symptoms; and sexual problems, bladder problems, and vaginal dryness, classified as urogenital symptoms [27].

Studies were excluded if they were duplicated reports, the duration of the study was less than 12 weeks, RCIE was combined with other plant medicines, lacked sufficient information, and if results were presented as graphics or percentage changes.

2.3. Data Extraction

The data were collected by the main author and then checked by the co-authors for correctness. The extracted data included the name of the first author; year of publication; country of origin; study design; follow-up period of the study; number of participants (randomized/analyzed); age (range) of women; daily dose of RCIE in the active arm (aglycone equivalent); a clearly described isoflavone component and their daily doses; baseline and final frequencies of HFs per day; scores of menopausal symptoms (without distinction between types of symptoms) or their differences; standard deviation, standard error, or 95% confidence intervals; and group size in each test arm.

2.4. Quality Assessment and Bias Risk of the Trials

The Cochrane risk of bias tool consists of seven items, which have a potential biasing influence on the estimates of an interventions' effectiveness in randomized trials: selection bias, performance bias, detection bias, attrition bias, reporting bias, and a catch-all item called "other sources of bias". The risk of bias in RCTs (randomized controlled trials) are included in the review as "High risk", "Unclear", or "Low risk" [28].

2.5. Statistical Analysis

The meta-analysis included all intervention groups from multi-arm studies. Moreover, to avoid duplication of data from the same people in surveys covering multiple time points, only one of such points was taken into account. The outcome measures were the difference in mean (MD) frequency of HFs between baseline and the end of the treatment, for both the intervention and control groups. Data of the size of the effects of RCIE in each study were presented as number of subjects (n) and the mean $\pm$ standard deviation (SD) of the differences. SD of MD was calculated using the following formula: SD = sqrt ((SD "baseline") 2 + (SD "endpoint") 2 − (2R × SD "baseline" × SD "endpoint")), where R is the correlation coefficient. We assumed an R of 0.40 to impute the missing SD of the mean within-group change according to Follman et al. [29]. If a 95% confidence interval (95% CI) was available for the difference in means, the same standard deviation was converted as: SD = sqrt (N) × (upper limit − lower limit)/(2u) (equal to 3.96). If the sample size was small (<30 in each group), the u-value could be obtained from tables of the t distribution with degrees of freedom equal to the sample size minus 1. This calculation is appropriate for data that are at least approximately normally distributed [30]. The random effects model was used to calculate the weighted mean difference (WMD) and 95% CI, and $p < 0.05$ was considered statistically significant. The results were assessed by comparing the mean $\pm$ SD of the change in HFs of the active group with the control group [31]. Cochrane Q and I^2 statistic were used to assess the heterogeneity. The I^2 test assessed whether the variance across studies was correct and not a result of a sampling error. The percentage of total variation indicated the degree of heterogeneity; I^2 values of $\leq 25\%$ were considered low, $>25\%$ as moderate, and $\geq 75\%$ as high heterogeneity [32]. STATISTICA Medical Software 11.0 StatSoft Poland, Krakow, Poland has been used for all statistical analyses.

3. Results

As a result of the search of electronic databases, 107 RCTs were identified. Sixty nine studies were excluded on the basis of title and/or abstracts. In the second phase, thirty eight potentially significant randomized controlled trials were identified and submitted for full-text assessment. Of these, twenty six studies were excluded due to failure to meet inclusion criteria. As a result, twelve RCTs that described the administration of RCIE to women for the management of HFs were included in the meta-analysis [33–44]. A detailed review of the selection procedure is shown in Figure 1.

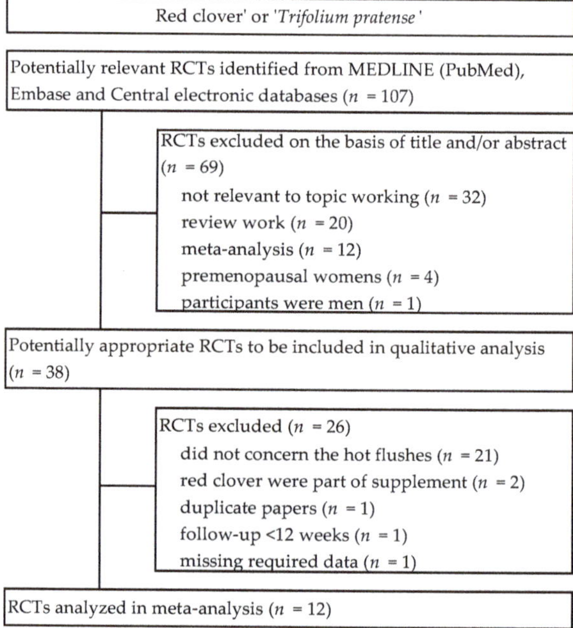

Figure 1. Flowchart of the selection procedure for studies included in the current review regarding red clover in menopausal symptoms. Abbreviations: RCTs, randomized controlled trials.

3.1. Characteristics of Included Trials

The characteristic of the selected twelve randomized, placebo-controlled clinical studies are reported in Table 1. All trials used a parallel group design, with the exception of three studies that used a crossover design [34,39,41]. The trials' duration ranged from 12 weeks to 2 years. Clinical studies were conducted in Australia (3), Peru, the Netherlands, the United States, the United Kingdom, Ecuador, Brazil, Austria, Iran, and Denmark. Overall, 1179 women experiencing menopause participated in the studies, and sample size ranged from 37 to 252 (1043 participants were included in the final analysis). Eight trials included postmenopausal women exclusively, three studies included women in both the peri- and postmenopausal period [37,38,40], and perimenopausal women were included in one study [44]. The average red clover isoflavone dose was 65.1 mg/d of aglycone equivalent (range, 37.1–160 mg/d). Two studies included two therapeutic arms with different doses of isoflavones [33,37]. The composition of the isoflavones and their doses varied among studies, which is shown in Table 1. Eight studies measured the daily frequency of HFs; the baseline of hot flushes was over three per day. Ten studies included the presence and/or severity of various somatic and psychological symptoms on scales to assess menopausal symptoms.

3.2. Assessment of the Methodological Quality of Trials

Details of the risk of bias assessment are shown in Figures 2 and 3. Several trials were characterized as "unclear risk", relating to the lack of sufficient information in the categories random sequence generation (selection bias) [34,35,41] (25%) and allocation concealment (selection bias) [34,35] (17%). The categories that presented a low risk of bias in all evaluated trials were the blinding of participants and personnel (performance bias) and the blinding of outcome assessment (detection bias). In the category of incomplete outcome data (attrition bias), "unclear bias" was demonstrated in 25% of studies [34,36,39]; it was not clear whether dropouts were likely to influence results. With respect to the selective reporting category, five studies [33–36,39] (42%) presented a "high risk of bias" associated with the lack of reports of adverse effects. "Unclear risk" in the other bias category was associated with an insufficient description of the study funding.

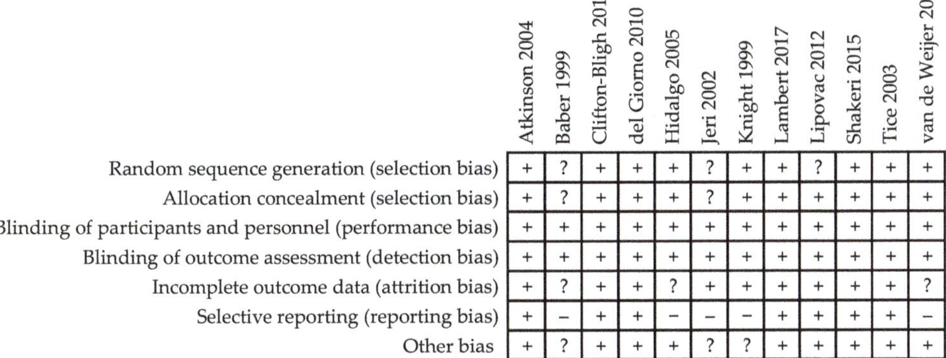

	Atkinson 2004	Baber 1999	Clifton-Bligh 2015	del Giorno 2010	Hidalgo 2005	Jeri 2002	Knight 1999	Lambert 2017	Lipovac 2012	Shakeri 2015	Tice 2003	van de Weijer 2002
Random sequence generation (selection bias)	+	?	+	+	+	?	+	+	?	+	+	+
Allocation concealment (selection bias)	+	?	+	+	+	?	+	+	+	+	+	+
Blinding of participants and personnel (performance bias)	+	+	+	+	+	+	+	+	+	+	+	+
Blinding of outcome assessment (detection bias)	+	+	+	+	+	+	+	+	+	+	+	+
Incomplete outcome data (attrition bias)	+	?	+	+	?	+	+	+	+	+	+	?
Selective reporting (reporting bias)	+	−	+	+	−	−	−	+	+	+	+	−
Other bias	+	?	+	+	+	?	?	+	+	+	+	+

+', low risk bias; '−', high risk of bias; '?', unknown bias

Figure 2. Risk of bias summary for each study as assessed by the authors [33–44].

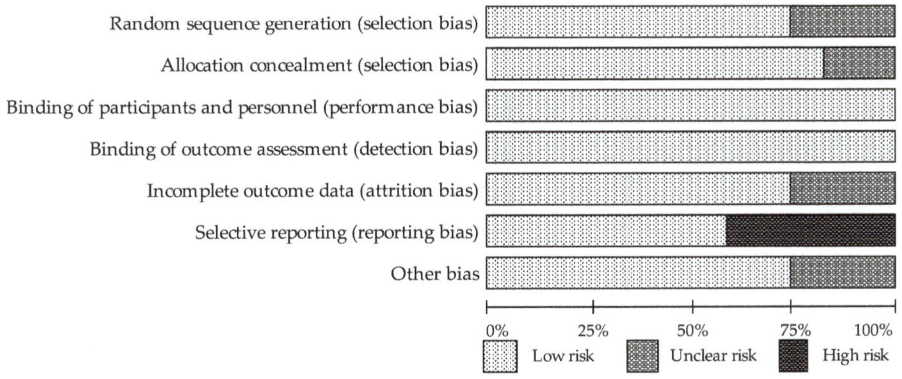

Figure 3. The assessment of risk of bias for each item; data are shown as a percentage for trials.

Table 1. Randomized controlled trials of Trifolium pratense for alleviating menopause symptoms: studies' characteristics.

First Author Pub. Data (Ref.) Country	Design Follow-up Period	Sample Size: Randomized/Analyzed	Participants Age, y (Range) Trial Inclusion Criteria	Intervention: Isoflavone Daily Dose	Baseline Hot Flush Frequency/d	Baseline Menopausal Score
Knight 1999 [33] Australia	Placebo controlled 3-arm parallel trial 1 wk placebo run-in/ 12 wk follow-up	37/37	54.6 ± 3.6 (40–65) Healthy postmenopausal women, bilateral oophorectomy or amenorrhea ≥ 6 mo, FSH > 40 mIU/mL, HF > 3/d	RCG "a"; 160 mg [a] RCG "b": 40 mg [b] PG: placebo	RCG "a": 9.0 ± 5.2 RCG "b": 6.9 ± 2.1 PG: 8.6 ± 4.6	GCS RCG "a": 19.9 ± 4.4 RCG "b": 19.9 ± 10.6 PG: 18.5 ± 11.4
Baber 1999 [34] Australia	Placebo controlled crossover trial 90 d active phase/ 7 d washout	51/51	54.0 ± 4.1 (45–65) Healthy postmenopausal women, age of menopause 50.0 ± 3.6 y, FSH > 30 mIU/mL, HF > 3/d	RCG: 40 mg [b] PG: placebo	RCG: 6.2 ± 2.7 PG: 6.4 ± 2.6	GCS RCG: 10.9 ± 6.5 PG: 12.3 ± 9.0
Jeri 2002 [35] Peru	Placebo controlled parallel trial 16 wk follow-up	30/30	51.0 ± 3.5 (<60) Healthy postmenopausal women, amenorrhea ≥ 12 mo, FSH > 30 mIU/mL, HF ≥ 5/d	RCG: 40 mg [b] PG: placebo	RCG: 7.0 ± 1.9 PG: 5.7 ± 1.6	- -
van de Weijer 2002 [36] Netherlands	Placebo controlled parallel trial 4 wk placebo run-in/ 12 wk follow-up	30/26	53.4 ± 6.3 (49–65) Healthy postmenopausal women, amenorrhea ≥ 12 mo, BMI 26.1 ± 4.2, HF ≥ 5/d	RCG: 80 mg [c] PG: placebo	RCG: 5.43 ± 2.6 PG: 5.6 ± 5.0	GCS RCG: 12.5 ± 11.2 PG: 13.8 ± 9.5
Tice 2003 [37] United States	Placebo controlled 3-arm parallel trial 2 wk placebo run-in/ 12 wk follow-up	252/252	52.3 ± 3.1 (45–60) Healthy peri- and post-menopausal women, 3.3 ± 4.5 ysm, FSH > 30 mIU/mL, BMI 26.1 ± 4.9, HF ≥ 35/wk	RCG "a": 80 mg [c] RCG "b": 57 mg [d] PG: placebo	RCG "a": 8.5 ± 5.8 RCG "b": 8.1 ± 3.0 PG: 7.8 ± 2.4	- - -
Atkinson 2004 [38] United Kingdom	Placebo controlled parallel trial 12 mo follow-up	205/99	52.2 ± 4.8 (49–65) Healthy peri- and post-menopausal women, FSH > 30 mIU/mL, BMI 25.3 ± 3.7, HF > 3/d	RCG: 40 mg [e] PG: placebo	RCG: 2.1 ± 2.7 PG: 2.5 ± 3.0	GCS RCG: 4.3 ± 4.3 PG: 4.3 ± 4.3

Table 1. *Cont.*

First Author Pub. Data (Ref.) Country	Design Follow-up Period	Sample Size: Randomized/Analyzed	Participants Age, y (Range) Trial Inclusion Criteria	Intervention: Isoflavone Daily Dose	Baseline Hot Flush Frequency/d	Baseline Menopausal Score
Hidalgo 2005 [39] Ecuador	Placebo controlled crossover trial 90 d active phase/ 7 d washout	60/53	51.3 ± 3.5 (>40) Healthy postmenopausal women, amenorrhea ≥ 12 mo, FSH > 35 mIU/ml, BMI 26.6 ± 3.9, KMI score ≥ 15	RCG: 80 mg [c] PG: placebo	- -	KMI RCG: 27.2 ± 7.7 PG: 27.2 ± 7.7
del Giorno 2010 [40] Brazil	Placebo controlled parallel trial 12 mo follow-up	120/100	55.5 ± 4.9 (45–65) Healthy peri- and post-menopausal women, amenorrhea ≥ 12 mo, FSH > 30 mIU/mL, BMI 28.8 ± 5.4	RCG: 40 mg PG: placebo	- -	KMI RCG: 25.3 ± 10.2 PG: 25.1 ± 9.0
Lipovac 2012 [41] Austria	Placebo controlled crossover trial 90 d active phase/ 7 d washout	113/109	54.1 ± 7.0 (>40) Healthy postmenopausal women, amenorrhea ≥ 12 mo, FSH > 35 mIU/mL, BMI 24.7 ± 3.9, HF > 5/d, KMI score ≥ 15/wk	RCG: 80 mg [c] CG: placebo	RCG: 11.7 ± 4.8 PG: 11.0 ± 5.1	KMIRCG: 32.5 ± 10.0 PG: 34.3 ± 10.4
Clifton-Bligh 2015 [42] Australia	Placebo controlled parallel trial 1 mo placebo run-in/ 2 y follow-up	147/103	54.4 ± 3.9 Healthy postmenopausal women, amenorrhea ≥ 12 mo, FSH > 30 mIU/mL, BMI 24.8 ± 4.3	RCG: 57 mg [d] PG: placebo	- -	GCS RCG: 8.9 ± 7.3 PG: 11.0 ± 8.0
Shakeri 2015 [43] Iran	Placebo controlled parallel trial 12 wk follow-up	72/71	54.8 ± 2.8 (50–59) Healthy postmenopausal women, 1.85 ± 0.9 ysm, BMD 21.1 ± 1.9	RCG: 80 mg [c] PG: placebo	- -	MRS RCG: 20.4 ± 6.3 PG: 20.8 ± 6.2
Lambert 2017 [44] Denmark	Placebo controlled parallel trial 12 wk follow-up	62/59	52.5 ± 3.5 (40–65) Healthy perimenopausal women, FSH ≥ 35 mIU/mL, BMI 25.7 ± 4.3, HF > 5/d	RCG: 37.1 mg [f] CG: placebo	RCG: 9.5 ± 6.4 PG: 8.6 ± 6.9	GCS RCG: 18.6 ± 12.3 PG: 20.8 ± 2.3

Data are presented as mean ± standard deviation. Abbreviations: -, data not available; BIO, biochanin A; BMI, body mass index (kg/m^2); DAI, daidzein; FOR, formononetin; FSH, follicle-stimulating hormone; GCS, Greene Climacteric Scale; GEN, genistein; GLY, glycitein; HF, hot flushes; KMI, Kupperman Menopausal Index; MRS, Menopause Rating Scale; PG, placebo group; pub. data, publication data; RCG, red clover group; ref., reference; VAS, vasomotor symptoms; ysm, years since menopause; mo, months; wk, week; d, day. Composition of isoflavones (aglycone, mg): [a] BIO (98.0), FOR (32.0), GEN (16.0), DAI (14.0); [b] BIO (24.5), FOR (8.0), GEN (4.0), DAI (3.5); [c] BIO (49.0), FOR (16.0), GEN (8.0), DAI (7.0); [d] FOR (44.6), BIO (5.8), DAI (1.8), GEN (0.8), GLY (0.8); [e] BIO (26.0), FOR (16.0), GEN (1.0), DAI (0.5); [f] FOR (19.0), BIO (9.0), GEN (4.2), DAI (1.6).

3.3. Systematic Review and Meta-Analysis

We present the results of a comprehensive systematic review with a meta-analysis regarding the impact assessment of RCIE on the incidence of hot flushes and on the presence and/or severity of various somatic and psychological symptoms in the evaluation of the menopausal symptom questionnaires and scales in perimenopausal and postmenopausal women.

3.3.1. Daily Hot Flushes Frequency

In most studies, a dose of 40–80 mg/d RCIE was used, except for the study of Lambert et al. [44], in which 37.1 mg/d was administered. In two trials, parallel comparisons with other doses were also performed: 57 mg/d [33] and 180 mg/d [37].

Of the eight RCTs with ten comparisons assessing frequency of HFs, six [33,35–37,41,44] showed a reduction in HFs, including four significant decreases [35,36,41,44] in the isoflavone group compared to the placebo group; while in four comparisons, no differences were observed between the groups [33,34,37,38].

A meta-analysis of all comparisons showed a statistically significant reduction in the daily incidence of HFs in women receiving active treatment compared to those receiving placebo treatment: WMD −1.73 HFs/d, 95% CI −3.28 to −0.18; p = 0.0292 (Figure 4). Additionally, a subgroup analysis was conducted to explain the possible influence of covariates on the observed high heterogeneity of included trials (I^2 = 87.34%) based on five prognostic factors: menopausal status, observation time, frequency of daily HFs, the total dose of isoflavones in terms of aglycone equivalents, and the differences in the types of isoflavone. Results of the sub-analysis are shown in Table 2. Differences in means were larger in comparisons that used RCIE at a dose of ≥80 mg/day (WMD −2.80 HFs/d; p = 0.1210), as well as when the formulations contained a higher proportion of biochanin A (−1.79 HFs/d; p = 0.0520), in postmenopausal women (WMD −2.68 HFs/d; p = 0.0105), with ≥5 HFs per day (WMD −2.56; p = 0.0096), and with an observation period of 12 weeks (WMD −1.95 HFs/d; p = 0.0206).

Study	RCE			Control			WMD (random)		WMD (random) 95% CI		Weight
	DM	SD	N	DM	SD	N	95% CI	p	Favours RCE	Favours control	%
Atkinson, 2004	−0.8	2.1	45	−1.1	1.8	54	0.20 (−0.58, 0.98)	0.6149			13,29
Baber. 1999	−1.37	17.66	25	−2.47	14.57	26	1.10 (−7.80, 10.00)	0.8087			2.49
Jeri, 2002	−3.4	1.83	15	−0.6	1.53	15	−2.80 (−4.01, −1.59)	0.0000			12.69
Knight, 1999 (160)	−3.1	5.39	13	−2.8	4.98	12	−0.30 (−4.36, 3.76)	0.8850			7.08
Knight, 1999 (40)	−2.0	4.4	12	−2.8	4.90	12	0.80 (−2.96, 4.56)	0.6767			7.61
Lambert, 2017	−2.97	4.74	30	0.04	6.55	29	−3.01 (−5.94, −0.05)	0.0438			9.24
Lipovac, 2012	−8.6	4.64	50	−0.9	7.47	50	−7.70 (−9.66, −5.74)	0.0000			11.27
Tice, 2003 (80)	−3.4	4.99	84	−2.8	3.32	85	−0.60 (−1.88, 0.68)	0.3567			12.58
Tice, 2003 (57)	−2.7	3.92	83	−2.8	3.42	85	0.10 (−1.00, 1.20)	0.8585			12.86
van de Weijer, 2002	−2.08	3.09	15	0.29	3.71	11	−2.37 (−6.12, 1.38)	0.0306			10.89
Total (95% CI)			**372**			**379**	**−1.73 (−3.28, −0.18)**	**0.0292**			**100.00**

Hetrogeneity: Q = 71.1148; df = 9 (p = 0.0000); Tau^2 = 4.5675; I^2 = 87.34%

Figure 4. Effects of isoflavones with red clover (*Trifolium pratense*) vs. placebo on the daily frequency of hot flushes in peri- and post-menopausal women. Number in brackets following author's name refers to dose of isoflavones in the study with more than one active group [33–38,41,44]. Abbreviations: RCIE, red clover isoflavone extract; WMD, weighted mean difference.

Table 2. Assessment of the effect of red clover isoflavones on the frequency of hot flushes in subgroup analysis.

Variables	*n*	Simple Size	WMD (95% CI)	*p*	I^2 (%)
Overall effects	10	751	−1.73 (−3.28, −0.18)	0.0292	87.34
Menopausal status					
Postmenopausal	7	315	−2.68 (−4.72, −0.63)	0.0105	71.44
Peri- and post-menopausal	3	436	0.01 (−0.55, 0.58)	0.9594	0.00
Follow-up period					
12 weeks	9	652	−1.95 (−3.61, −0.30)	0.0206	81.33
12 months	1	99	0.20 (−0.58, 0.98)	0.6149	(-)
Frequency of hot flushes					
≥5/day	6	552	−2.56 (−4.49, −0.62)	0.0096	87.67
≥3/day	4	199	0.21 (−0.53, 0.96)	0.5761	0.00
Isoflavones dose					
<80 mg/day	6	431	−0.88 (−2.34, 0.58)	0.2370	76.83
≥80 mg/day	4	320	−2.80 (−6.35, 0.74)	0.1210	86.61
Dominant of isoflavones					
Biochanin A	8	524	−1.79 (−3.60, 0.02)	0.0520	85.78
Formononetin	2	227	−1.14 (−4.13, 1.84)	0.4519	73.64

n, number of comparisons; -, not calculated.

3.3.2. Rating of Menopausal Complaints Using Instruments to Measure Intensity of Symptoms

In three studies [39–41], the Kupperman Menopausal Index (KMI) scale was used to assess the severity of climacteric symptoms. The results of the meta-analysis indicated that compared to placebo, RCIE was effective in relieving menopausal symptoms: WMD −12.77, 95% CI −23.61 to −1.93; p = 0.0209; I^2 = 96.19% (Figure 5A). Of the six RCTs with seven comparisons that reported Greene Climacteric Scale (GCS) data, four (40 mg/d) [33,34,37,38] showed a marginal, insignificant decrease in GCS scores, in contrary to three studies (160 mg/d) [33,42,44] that reported a slight, insignificant intensification of menopausal symptoms. Meta-analysis did not show the beneficial effect of RCIE on menopausal symptoms and complaints in the GCS scale used in these studies, compared with the placebo: WMD 0.11, 95% CI −0.87 to 1.09; p = 0.8265; I^2 = 0.00% (Figure 5B).

According to the results obtained from a single RCT [43], women receiving RCIE observed a significant reduction in their Menopause Rating Scale (MRS) score compared to placebo: WMD −6.81, 95% CI −9.79 to −3.83; p = 0.0000 (Figure 5C).

A. Kupperman Menopausal Index

Study	Intervention			Control			WMD (random)	WMD (random) 95% CI	Weight
	MD	SD	N	MD	SD	N	95% CI	Favours RCE — Favours control	%
Hidalgo, 2005	−21.3	7.1	53	−6.3	9.4	53	−15.0 (−17.76, −12.24)		33.96
del Giorno, 2010	−14.22	10.87	50	−13.11	9.88	50	−1.11 (−5.18, 2.96)		33.10
Lipovac, 2012	−24.5	9.61	50	−2.3	13.14	59	−22.2 (−26.48, −17.92)		32.94
Total (95% CI)			**153**			**162**	**−12.77 (−23.61, −1.93)**		**100.0**
							p = **0.0209**		

Hetrogeneity: Q = 52.5345; df = 2 (*p* = 0.0000); Tau^2 = 88.0997; I2 = 96.19%

B. Greene Climacteric Scale

Study	Intervention			Control			WMD (random)	WMD (random) 95% CI	Weight
	MD	SD	N	MD	SD	N	95% CI	Favours RCE — Favours control	%
Knight, 1999 (40)	−8.7	10.73	12	−8.6	10.53	12	3.40 (−8.95, 13.75)		0.89
Knight, 1999 (160)	−5.2	15.57	13	−8.6	10.53	12	−0.10 (−8.61, 8.41)		1.32
Baber, 1999	−4.33	6.39	25	−4.04	9.59	26	−0.29 (−4.75, 4.17)		4.83
van de Weijer, 2002	−1.6	11.6	15	0.8	11.82	11	−2.40 (−11.52, 6.72)		1.15
Atkinson, 2004	−0.9	4.1	70	−0.8	2.8	83	−0.10 (−1.23, 1.03)		74.57
Clifton-Bligh, 2015	1.313	7.41	56	0.099	7.15	47	1.21 (−1.60, 4.03)		12.06
Lambert, 2017	−3.85	8.12	30	−4.82	8.73	29	0.97 (−3.34, 5.28)		5.17
Total (95% CI)			**221**			**220**	**0.11 (−0.87, 1.09)**		**100.0**
							p = **0.8265**		

Hetrogeneity: Q = 1.5865; df = 6 (*p* = 0.9535); Tau^2 = 0.0000; I2 = 0.00%

C. Menopause Rating Scale

Study	Intervention			Control			WMD (random)	WMD (random) 95% CI	Weight
	MD	SD	N	MD	SD	N	95% CI	Favours RCE — Favours control	%
Shakeri, 2015	−10.33	5.8	35	−3.52	6.99	36	−6.81 (−9.79, −3.83)		100.0
Total (95% CI)			**35**			**36**	**−6.81 (−9.79, −3.83)**		**100.0**
							p = **0.0000**		

Figure 5. Effects of red clover (*Trifolium pratense*) isoflavones vs. placebo on rating menopausal symptoms using the following questionnaires, based on the respondents' replies concerning the intensity of complaints. Number in brackets following author's name refers to the dose of isoflavones in the study with more than one active group [33,34,36,38–44]. The letter A marks the first part of the figure containing the Kupperman Menopausal Index analysis. The letter B marks the second part of the figure containing the Greene Climacteric Scale analysis. The letter C marks the third part of the figure containing the Menopause Rating Scale analysis. Abbreviations: WMD, weighted mean difference.

4. Discussion

Our meta-analysis of all comparisons showed a statistically significant reduction in the daily incidence of HFs in women receiving active treatment compared to those receiving placebo, WMD −1.73, 95% CI −3.28 to 0.18; *p* = 0.0292. These results are consistent with those presented in previous meta-analyses [45–47] that also found some beneficial effects of RCIE, although not always statistically significant. Myers and Vigar [45], based on the analysis of 5 studies [33,36,37,39,41] including 438 women, showed a statistically significant reduction in the number of HFs after a daily intake of ≥80 mg RCIE compared to placebo: WMD −3.46, 95% CI −4.37 to −2.56; *p* < 0.00001. Coon et al. [46] reported a significant decrease in the daily episodes of flushing (WMD −1.63, 95% CI −2.97 to −0.28; *p* < 0.02) in an analysis that included 5 trials [33–37] with 7 comparisons and 385 participants; the doses of RCI were 40, 57, 80–82, or 160 mg/d. In turn, a meta-analysis by Ghazanfarpour et al. [47] based on 6 studies [33–37,41] showed a decrease in HFs/day frequency, close to statistical significance, post-administration of RCIE in 40 or 80 mg: WMD −1.99, 95% CI −4.12 to 0.19; p = 0.067. Furthermore, three meta-analyses [48–50] showed positive effects of RCIE, but they were statistically non-significant on the frequency of HFs. Lethaby et al. [48], using five studies [33–37], showed a slight decrease, WMD −0.93, 95% CI −1.95 to 0.10; p = 0.21 (40 mg/d, 80 mg/d doses). In turn, Nelson et al. [49] reported a reduction in HFs (WMD = −0.44, 95% CI −1.47 to 0.58) based on 9 comparisons (6 studies [33–38]; doses of 40 mg/d, 57 mg/d, 80–82 mg/d, 160 mg/d). While Franco et al. [50] noted a decrease in the number of daily hot flushes for red clover isoflavones compared with placebo: WMD = −1.84, 95% −3.87 to 0.19; p = 0.20 (7 studies [33–38,41]; doses: 40–160 mg).

The discussion omitted the results of systematic reviews with a meta-analysis [51,52] with the adopted research methodology based on the analysis of only the final values of HFs at the end of treatment.

Our meta-analysis also showed that RCIE significantly lowered the KMI points— WMD −12.77 (p = 0.0209) and the MRS points—WMD −6.81 (p = 0.0000). The last result was only based on one study [43], which makes it impossible to draw a final conclusion. The earlier meta-analysis of Myers and Vigor [45] assessing the effect of RCIE on menopausal symptoms on the KMI scale, based on two trials [39,41], also demonstrated a marked significant reduction of WMD −21.8 points (p < 0.00001) in women receiving red clover. It is important to note that the lack of a significant reduction on the GCS (WMD 0.11 points; p = 0.8265) found in our analysis may undermine the usefulness of RCIE in relieving menopausal symptoms other than HFs in perimenopausal and postmenopausal women.

A number of possible limitations should be taken into account when interpreting the results of the present meta-analysis. The fact that most of these assessments were based on a relatively limited number of available trials as well as the small number of participants in some studies may result in insufficient statistical power and limit the draw-down of definitive conclusions. Also, inter-individual differences in the metabolism and bioavailability of isoflavones may cause variability in the response to their use. The placebo effect should also be taken into account: clear placebo responses in menopausal women in drug trials affecting vasomotor symptoms are well documented [53]. Furthermore, the actual dose of aglycone isoflavones administered or absorbed was difficult to determine. Other limitations are related to the fact that the analyzed works may not represent all research related to this topic, especially articles published in languages other than English. Additionally, if the error of published research is strong, it is possible to have overestimated or underestimated the effect of red clover on menopausal symptoms.

5. Conclusions

This meta-analysis of randomized controlled trials assessing the effect of a specific standardized extract of red clover isoflavones on menopausal symptoms showed a statistically moderate relationship with the reduction in the daily frequency of hot flushes. However, further well-designed studies are required to confirm the present findings and to finally determine the effects of red clover on the relief of flushing episodes, to provide more comprehensive information about well-defined preparations, and the optimal dose and duration of taking red clover aglycones to achieve their highest effectiveness.

Author Contributions: Conceptualization, W.K., A.B. (Agnieszka Barańska); Data curation, A.B. (Agata Błaszczuk), M.P.-D.; Formal analysis, M.M.; Funding acquisition, Bartłomiej Drop; Investigation, A.B. (Agata Błaszczuk); Methodology, W.K., A.B. (Agnieszka Barańska), and M.M.; Project administration, M.P.-D.; Resources, A.B. (Agnieszka Barańska); Software, W.K.; Supervision, Krzysztof Kanecki; Validation, W.K.; Visualization, A.B. (Agata Błaszczuk); Writing—original draft, W.K.; Writing—review and editing, A.B. (Agnieszka Barańska). All authors have read and agreed to the published version of the manuscript.

Funding: This research received no external funding.

Institutional Review Board Statement: Not applicable.

Informed Consent Statement: Not applicable.

Data Availability Statement: Not applicable.

Acknowledgments: Not applicable.

Conflicts of Interest: The authors declare no conflict of interest.

References

1. Thurston, R.C. Vasomotor symptoms: Natural history, physiology, and links with cardiovascular health. *Climacteric* **2018**, *21*, 96–100. [CrossRef]
2. Monteleone, P.; Mascagni, G.; Giannini, A.; Genazzani, A.R.; Simoncini, T. Symptoms of menopause—Global prevalence, physiology and implications. *Nat. Rev. Endocrinol.* **2018**, *14*, 199–215. [CrossRef]
3. Avis, N.E.; Crawford, S.L.; Greendale, G.; Bromberger, J.T.; Everson-Rose, S.A.; Gold, E.B.; Hess, R.; Joffe, H.; Kravitz, H.M.; Tepper, P.G.; et al. Duration of menopausal vasomotor symptoms over the menopause transition. *JAMA Intern. Med.* **2015**, *175*, 531–539. [CrossRef]
4. Freeman, E.W.; Sherif, K. Prevalence of hot flushes and night sweats around the world: A systematic review. *Climacteric* **2007**, *10*, 197–214. [CrossRef]
5. Gold, E.B.; Colvin, A.; Avis, N.; Bromberger, J.; Greendale, G.A.; Powell, L.; Sternfeld, B.; Matthews, K. Longitudinal analysis of the association between vasomotor symptoms and race/ethnicity across the menopausal transition: Study of women's health across the nation. *Am. J. Public Health* **2006**, *96*, 1226–1235. [CrossRef]
6. Duffy, O.K.; Iversen, L.; Hannaford, P.C. Factors associated with reporting classic menopausal symptoms differ. *Climacteric* **2013**, *16*, 240–251. [CrossRef] [PubMed]
7. Herber-Gast, G.-C.M.; Mishra, G.D.; van der Schouw, Y.T.; Brown, W.J.; Dobson, A.J. Risk factors for night sweats and hot flushes in midlife: Results from a prospective cohort study. *Menopause* **2013**, *20*, 953–959. [CrossRef] [PubMed]
8. Chung, H.-F.; Pandeya, N.; Dobson, A.J.; Kuh, D.; Brunner, E.J.; Crawford, S.L.; Avis, N.E.; Gold, E.B.; Mitchell, E.S.; Woods, N.F.; et al. The role of sleep difficulties in the vasomotor menopausal symptoms and depressed mood relationships: An international pooled analysis of eight studies in the InterLACE consortium. *Psychol. Med.* **2018**, *48*, 2550–2561. [CrossRef] [PubMed]
9. Geukes, M.; van Aalst, M.P.; Robroek, S.J.; Laven, J.S.E.; Oosterhof, H. The impact of menopause on work ability in women with severe menopausal symptoms. *Maturitas* **2016**, *90*, 3–8. [CrossRef] [PubMed]
10. Rossmanith, W.G.; Ruebberdt, W. What causes hot flushes? The neuroendocrine origin of vasomotor symptoms in the menopause. *Gynecol. Endocrinol.* **2009**, *25*, 303–314. [CrossRef] [PubMed]
11. Freedman, R.R. Menopausal hot flashes: Mechanisms, endocrinology, treatment. *J. Steroid Biochem. Mol. Biol.* **2014**, *142*, 115–120. [CrossRef] [PubMed]
12. Tao, M.; Teng, Y.; Shao, H.; Wu, P.; Mills, E.J. Knowledge, perceptions and information about hormone therapy (HT) among menopausal women: A systematic review and meta-synthesis. *PLoS ONE* **2011**, *6*, e24661. [CrossRef] [PubMed]
13. Marjoribanks, J.; Farquhar, C.; Roberts, H.; Lethaby, A.; Lee, J. Long-term hormone therapy for perimenopausal and post-menopausal women. *Cochrane Database Syst. Rev.* **2017**, *1*, CD004143. [CrossRef] [PubMed]
14. French, L.M.; Smith, M.A.; Holtrop, J.S.; Holmes-Rovner, M. Hormone therapy after the Women's Health Initiative: A qualitative study. *BMC Fam. Pract.* **2006**, *7*, 61. [CrossRef] [PubMed]
15. Schonberg, M.A.; Davis, R.B.; Wee, C.C. After the women's health initiative: Decision making and trust of women taking hormone therapy. *Women's Health Issues* **2005**, *15*, 187–195. [CrossRef]
16. Nedrow, A.; Miller, J.; Walker, M.; Nygren, P.; Huffman, L.H.; Nelson, H.D. Complementary and alternative therapies for the management of menopause-related symptoms: A systematic evidence review. *Arch. Intern. Med.* **2006**, *166*, 1453–1465. [CrossRef]
17. Wong, V.C.; Lim, C.E.; Luo, X.; Wong, W.S. Current alternative and complementary therapies used in menopause. *Gynecol. Endocrinol.* **2009**, *25*, 166–174. [CrossRef]
18. Mintziori, G.; Lambrinoudaki, I.; Goulis, D.G.; Ceausu, I.; Depypere, H.; Erel, C.T.; Pérez-López, F.R.; Schenck-Gustafsson, K.; Simoncini, T.; Tremollieres, F.; et al. EMAS position statement: Non-hormonal management of menopausal vasomotor symptoms. *Maturitas* **2015**, *81*, 410–413. [CrossRef]
19. Booth, N.L.; Piersen, C.E.; Banuvar, S.; Geller, S.E.; Shulman, L.P.; Farnsworth, N.R. Clinical studies of red clover (*Trifolium pratense*) dietary supplements in menopause: A literature review. *Menopause* **2006**, *13*, 251–264. [CrossRef]
20. Lemežienė, N.; Padarauskas, A.; Butkutė, B.; Cesevičienė, J.; Taujenis, L.; Norkevičienė, E. The concentration of isoflavones in red clover (*Trifolium pratense* L.) at flowering stage. *Zemdirb.-Agric.* **2015**, *102*, 443–448. [CrossRef]
21. Tolleson, W.H.; Doerge, D.R.; Churchwell, M.I.; Marques, M.M.; Roberts, D.W. Metabolism of biochanin A and formononetin by human liver microsomes in vitro. *J. Agric. Food Chem.* **2002**, *50*, 4783–4790. [CrossRef]
22. Liu, J.; Burdette, J.E.; Xu, H.; Gu, C.; van Breemen, R.B.; Bhat, K.P.; Booth, N.; Constantinou, A.I.; Pezzuto, J.M.; Fong, H.H.; et al. Evaluation of estrogenic activity of plant extracts for the potential treatment of menopausal symptoms. *J. Agric. Food Chem.* **2001**, *49*, 2472–2479. [CrossRef]

23. Morito, K.; Aomori, T.; Hirose, T.; Kinjo, J.; Hasegawa, J.; Ogawa, S.; Inoue, S.; Muramatsu, M.; Masamune, Y. Interaction of phy-toestrogens with estrogen receptors alpha and beta (II). *Biol. Pharm. Bull.* **2002**, *25*, 48–52. [CrossRef]

24. Welch, V.; Petticrew, M.; Petkovic, J.; Moher, D.; Waters, E.; White, H.; Tugwell, P.; Atun, R.; Awasthi, S.; Barbour, V.; et al. Extending the PRISMA statement to equity-focused systematic reviews (PRISMA-E 2012): Explanation and elaboration. *J. Clin. Epidemiol.* **2016**, *70*, 68–89. [CrossRef] [PubMed]

25. Kupperman, H.S.; Blatt, M.H.; Wiesbader, H.; Filler, W. Comparative clinical evaluation of estrogenic preparations by the meno-pausal and amenorrheal indices. *J. Clin. Endocrinol. Metab.* **1953**, *13*, 688–703. [CrossRef]

26. Greene, J.G. A factor analytic study of climacteric symptoms. *J. Psychosom. Res.* **1976**, *20*, 425–430. [CrossRef]

27. Schneider, H.P.; Heinemann, L.A.; Rosemeier, H.P.; Potthoff, P.; Behre, H.M. The Menopause Rating Scale (MRS): Reliability of scores of menopausal complaints. *Climacteric* **2000**, *3*, 59–64. [CrossRef]

28. Higgins, J.P.; Altman, D.G.; Gøtzsche, P.C.; Jüni, P.; Moher, D.; Oxman, A.D.; Savović, J.; Schulz, K.F.; Weeks, L.; Sterne, J.A.C.; et al. The Cochrane Collaboration's tool for assessing risk of bias in randomized trials. *BMJ* **2011**, *343*, 5928. [CrossRef] [PubMed]

29. Follmann, O.; Elliott, P.; Suh, I.; Cutler, J. Variance imputation for overviews of clinical trials with continuous response. *J. Clin. Epidemiol.* **1999**, *45*, 769–773. [CrossRef]

30. Higgins, J.P.T.; Deeks, J.J. Selecting studies and collecting data. In *Cochrane Handbook for Systematic Reviews of Interventions*; Higgins, J.P.T., Green, S., Eds.; The Cochrane Collaboration; John Wiley & Sons: Chichester, UK, 2008; pp. 172–178.

31. DerSimonian, R.; Laird, N. Meta-analysis in clinical trials. *Control Clin. Trials* **1986**, *7*, 177–188. [CrossRef]

32. Higgins, J.P.; Thompson, S.G. Quantifying heterogeneity in a meta-analysis. *Stat. Med.* **2002**, *21*, 1539–1558. [CrossRef] [PubMed]

33. Knight, D.C.; Howes, J.B.; Eden, J.A. The effect of Promensil, an isoflavone extract, on menopausal symptoms. *Climacteric* **1999**, *2*, 79–84. [CrossRef]

34. Baber, R.J.; Templeman, C.; Morton, T.; Kelly, G.E.; West, L. Randomized placebo-controlled trial of an isoflavone supplement and menopausal symptoms in women. *Climacteric* **1999**, *2*, 85–92. [CrossRef] [PubMed]

35. Jeri, A.R. The use of an isoflavone supplement to relieve hot flashes. *Female Patient* **2002**, *27*, 35–37.

36. van de Weijer, P.H.; Barentsen, R. Isoflavones from red clover (Promensil) significantly reduce menopausal hot flush symptoms compared with placebo. *Maturitas* **2002**, *42*, 187–193. [CrossRef]

37. Tice, J.A.; Ettinger, B.; Ensrud, K.; Wallace, R.; Blackwell, T.; Cummings, S.R. Phytoestrogen supplements for the treatment of hot flashes: The isoflavone clover extract (ICE) study: A randomized controlled trial. *JAMA* **2003**, *290*, 207–214. [CrossRef]

38. Atkinson, C.; Warren, R.M.L.; Sala, E.; Dowsett, M.; Dunning, A.M.; Healey, C.S.; Runswick, S.; Day, N.E.; Bingham, S.A. Red clover-derived isoflavones and mammographic breast density: A double-blind, randomized, placebo-controlled trial [ISRCTN42940165]. *Breast Cancer Res.* **2004**, *6*, R170–R179. [CrossRef]

39. Hidalgo, L.A.; Chedraui, P.A.; Morocho, N.; Ross, S.; Miguel, G.S. The effect of red clover isoflavones on menopausal symptoms, lipids and vaginal cytology in menopausal women: A randomized, double-blind, placebo-controlled study. *Gynecol. Endocrinol.* **2005**, *21*, 257–264. [CrossRef]

40. del Giorno, C.; Fonseca, A.M.; Bagnoli, V.R.; Assis, J.S.; Soares, J.M., Jr.; Baracat, E.C. Effects of *Trifolium pratense* on the climacteric and sexual symptoms in postmenopause women. *Rev. Assoc. Med. Bras.* **2010**, *56*, 558–562.

41. Lipovac, M.; Chedraui, P.; Gruenhut, C.; Gocan, A.; Kurz, C.; Neuber, B.; Imhof, M. The effect of red clover isoflavone supplementation over vasomotor and menopausal symptoms in postmenopausal women. *Gynecol. Endocrinol.* **2012**, *28*, 203–207. [CrossRef]

42. Clifton-Bligh, P.B.; Nery, M.L.; Cliftonbligh, R.J.; Visvalingam, S.; Fulcher, G.R.; Byth, K.; Baber, R. Red clover isoflavones enriched with formononetin lower serum LDL cholesterol—A randomized, double-blind, placebo-controlled study. *Eur. J. Clin. Nutr.* **2015**, *69*, 134–142. [CrossRef]

43. Shakeri, F.; Taavoni, S.; Goushegir, A.; Haghani, H. Effectiveness of red clover in alleviating menopausal symptoms: A 12-week randomized, controlled trial. *Climacteric* **2015**, *18*, 568–573. [CrossRef]

44. Lambert, M.N.T.; Thorup, A.C.; Hansen, E.S.S.; Jeppesen, P.B. Combined red clover isoflavones and probiotics potently reduce menopausal vasomotor symptoms. *PLoS ONE* **2017**, *12*, e0176590. [CrossRef]

45. Myers, S.P.; Vigar, V. Effects of a standardised extract of *Trifolium pratense* (Promensil) at a dosage of 80 mg in the treatment of menopausal hot flushes: A systematic review and meta-analysis. *Phytomedicine* **2017**, *24*, 141–147. [CrossRef] [PubMed]

46. Coon, J.T.; Pittler, M.H.; Ernst, E. *Trifolium pratense* isoflavones in the treatment of menopausal hot flushes: A systematic review and meta-analysis. *Phytomedicine* **2007**, *14*, 153–159. [CrossRef] [PubMed]

47. Ghazanfarpour, M.; Sadeghi, R.; Roudsari, R.L.; Khorsand, I.; Khadivzadeh, T.; Muoio, B. Red clover for treatment of hot flashes and menopausal symptoms: A systematic review and meta-analysis. *J. Obstet. Gynaecol* **2016**, *36*, 301. [CrossRef] [PubMed]

48. Lethaby, A.; Marjoribanks, J.; Kronenberg, F.; Roberts, H.; Eden, J.; Brown, J. Phytoestrogens for menopausal vasomotor symptoms. *Cochrane Database Syst. Rev.* **2013**, *12*, CD001395. [CrossRef] [PubMed]

49. Nelson, H.D.; Vesco, K.K.; Haney, E.; Fu, R.; Nedrow, A.; Miller, J.; Nicolaidis, C.; Walker, M.; Humphrey, L. Nonhormonal therapies for menopausal hot flashes: Systematic review and meta-analysis. *JAMA* **2006**, *295*, 2057–2071. [CrossRef]

50. Franco, O.H.; Chowdhury, R.; Troup, J.; Voortman, T.; Kunutsor, S.; Kavousi, M.; Oliver-Williams, C.; Muka, T. Use of plant-based therapies and menopausal symptoms: A systematic review and meta-analysis. *JAMA* **2016**, *315*, 2554–2563. [CrossRef] [PubMed]

51. Krebs, E.E.; Ensrud, K.E.; Macdonald, R.; Wilt, T.J. Phytoestrogens for treatment of menopausal symptoms: A systematic review. *Obstet. Gynecol.* **2004**, *104*, 824–836. [CrossRef]

52. Gartoulla, P.; Han, M.M. Red clover extract for alleviating hot flushes in postmenopausal women: A meta-analysis. *Maturitas* **2014**, *79*, 58–64. [CrossRef] [PubMed]
53. Li, L.; Xu, L.; Wu, J.; Dong, L.; Lv, Y.; Zheng, Q. Quantitative analysis of placebo response and factors associated with menopausal hot flashes. *Menopause* **2017**, *24*, 932–937. [CrossRef] [PubMed]

Article

Associations between Diet Quality and Anthropometric Measures in White Postmenopausal Women

Andrea Y. Arikawa [1,*] and Mindy S. Kurzer [2]

1 Department of Nutrition and Dietetics, University of North Florida, Jacksonville, FL 32224, USA
2 Department of Food Science and Nutrition, University of Minnesota, Saint Paul, MN 55108, USA; mkurzer@umn.edu
* Correspondence: andrea.arikawa@unf.edu; Tel.: +1-904-479-8995

Abstract: The purpose of this cross-sectional study was to examine the relationship between diet and anthropometric measures in postmenopausal women. Data collected from 937 women enrolled in the Minnesota Green Tea Trial (NTC00917735) were used for this analysis. Dietary intake and health-related data were collected via questionnaires. Body weight, height, and waist circumference (WC) were measured by the study staff. The mean age of participants was 59.8 years and mean WC was 83 cm. Approximately 30% of the participants had WC greater than 88 cm. Healthy Eating Index-2015 score was 72.6 and the Dietary Inflammatory Index score was 0. Intakes of whole grains, dairy, protein, sodium, and saturated fat did not meet the dietary guidelines. Only 12.5% consumed the recommended daily amount of calcium (mean intake = 765 mg/day). When calcium supplements were considered, only 35.2% of the participants had adequate intakes, even though 68.9% reported taking a calcium supplement. We found that age and number of medications taken were significantly associated with waist circumference ($p = 0.005$). Women who reported taking two or more medications had greater WC (85 cm) compared to women who reported not taking any medications (82.2 cm), $p = 0.002$. Our findings suggest that achieving adequate calcium and vitamin D intake may be challenging to postmenopausal women.

Keywords: healthy eating index; calcium; waist circumference; diet history questionnaire; supplements; menopause; bone mineral density

Citation: Arikawa, A.Y.; Kurzer, M.S. Associations between Diet Quality and Anthropometric Measures in White Postmenopausal Women. *Nutrients* **2021**, *13*, 1947. https://doi.org/10.3390/nu13061947

Academic Editor: Masakazu Terauchi

Received: 7 May 2021
Accepted: 4 June 2021
Published: 6 June 2021

Publisher's Note: MDPI stays neutral with regard to jurisdictional claims in published maps and institutional affiliations.

Copyright: © 2021 by the authors. Licensee MDPI, Basel, Switzerland. This article is an open access article distributed under the terms and conditions of the Creative Commons Attribution (CC BY) license (https://creativecommons.org/licenses/by/4.0/).

1. Introduction

The transition to menopause is accompanied by several concomitant changes in women's biological, physiological, behavioral, and social characteristics that will define women's well-being and future risk of disease. One of the major physiological changes that occur with menopause is a marked decline in estradiol production, which is associated with several adverse consequences such as sleep disturbances, vasomotor symptoms such as hot flashes and night sweats, sexual dysfunction, metabolic disorders, bone loss, dyslipidemia, and obesity [1,2].

Approximately 75% of women ages 60 years and older are overweight [3]. Previous research indicates that fat mass increases during the years preceding the transition to menopause and abruptly worsens at the onset of menopause, followed by stabilization [4]. Although these observed increases in fat mass seem to be related to aging, rather than menopause, evidence suggests that central fat accumulation is higher in postmenopausal women compared with premenopausal women [3], which is attributed to the decreased levels of estradiol and consequent increases in the activity of androgen receptors in visceral adipose tissue [5]. In addition to changes in fat mass, a sharp menopause-related increase in low-density lipoprotein (LDL)-cholesterol has been reported in the cohort Study of Women's Health Across the Nation (SWAN) [6], which could predispose postmenopausal women to cardiovascular disease and other chronic diseases such as diabetes. Indeed, menopause has been recognized as a female-specific risk factor for cardiovascular disease by the American Heart Association [7].

Bone loss is considered a hallmark of menopause. A decline in bone mineral density during the transition to menopause has been well documented in prospective cohort studies [8,9]. The role of diet on the prevention of bone mineral loss is not entirely elucidated in light of conflicting findings regarding the benefits of calcium and vitamin D supplementation for the prevention of fractures [10–12]. Nonetheless, adequate intake of these two nutrients is recommended for the prevention of bone loss and fractures in postmenopausal women [13]. Adequate protein intake also seems to be associated with bone mineral density. Protein intake representing around 24% of total energy intake has been associated with slower rates of bone loss and lower risk of hip fractures in older individuals diagnosed with osteoporosis [14]. Finally, regular physical activity has also been shown to positively affect bone health in women [15].

While some of the natural changes that occur with menopause are not modifiable, lifestyle behaviors, particularly dietary behavior and physical activity, can be modified to reduce the impact of disease risk factors on women's health such as weight gain, dyslipidemia, and bone loss, among others. To target dietary behavior as a means of lowering the risk of disease in postmenopausal women, it is necessary to identify specific diet-related concerns in this population. Several measures of diet quality can be used to assess dietary patterns of postmenopausal women; the Healthy Eating Index (HEI) is a commonly used index of diet quality because it reflects dietary adequacy in relation to the Dietary Guidelines for Americans [16]; the Dietary Inflammatory Index (DII) assesses the inflammatory potential of the diet on a continuum from anti-inflammatory, represented by lower scores to pro-inflammatory, represented by higher scores [17]. These indices are useful in the context of large datasets because they provide a measure of overall dietary complexity, which can be analyzed in conjunction with intakes of individual nutrients to obtain a more thorough assessment of an individual's diet. The purpose of this study was to examine relationships between diet quality as assessed by the HEI and the DII and anthropometric measures in a large sample of postmenopausal women living in a Midwestern state of the U.S.

2. Materials and Methods

Baseline data collected from 937 postmenopausal women enrolled in the Minnesota Green Tea Trial (MGTT) were used for this cross-sectional analysis. The MGTT was a double-blinded placebo-controlled trial that assessed the effects of green tea catechin supplements on biomarkers of breast cancer risk (clinical trial ID: NCT00917735) [18].

2.1. Study Participants

Eligible participants had to be between the ages of 50 and 70, report being generally healthy, postmenopausal for at least one year, not using any hormone therapy within the past six months, and having been diagnosed with heterogeneously or extremely dense breasts on a recent screening mammogram. Those who reported using hormone replacement therapy, elevated liver enzymes, or over than 5 kg change in body weight over the past year were not eligible. Written consent was obtained from all participants and the trial protocol was approved by the University of Minnesota Institutional Review Board. Details about participant selection and enrollment into the MGTT have been published previously [18].

2.2. Dietary Assessment

Dietary intake was assessed using the National Cancer Institute's diet history questionnaire (DHQ-1) [19]. This questionnaire included 124 food items with details about portion size and frequency of consumption including vitamin and mineral supplements.

Data originated from the DHQ-1 was used to calculate the Healthy Eating Index (HEI)-2015, according to the method previously described [20]. Briefly, the HEI is a measure of diet quality based on alignment of dietary components to the recommendations of the

Dietary Guidelines for Americans [16]. The HEI includes 13 components from which the sum can result in scores ranging from 0 to 100. The higher the HEI score, the greater the alignment between the individuals' diets and dietary recommendations. Previous research has shown that the HEI is valid and reliable when tested using dietary data from a representative sample of Americans [21].

DHQ-1 data were also used to calculate the Dietary Inflammatory Index (DII) score, as described by Shivappa et al. [17]. The z-scores for intakes of 28 food parameters were calculated (calories (kcal), carbohydrate (g), protein (g), dietary fiber (g), saturated fat (g), trans fats (g), monounsaturated fat (g), omega-6 (g), omega-3 (g), cholesterol (mg), alcohol (g), vitamin A (µg), β-carotene (µg), thiamin (mg), niacin (mg), riboflavin (mg), vitamin B6 (mg), folate (µg), vitamin B12 (µg), vitamin C (mg), vitamin D (µg), vitamin E (mg), magnesium (mg), iron (mg), zinc (mg), selenium (mg), caffeine (mg), isoflavones (mg)). Percentile scores were calculated from the z-scores and then doubled and 'one' was subtracted from the values to achieve a symmetrical distribution. Next, each food parameter's centered percentile value was multiplied by an 'inflammatory effect score' and all scores for the 28 food parameters were summed to obtain the overall DII for each study participant. Intake of nutrients via supplements was not included in the calculation of the DII. Although the original DII includes 45 food parameters, several of these were not available from the DHQ-1. However, it can be argued that food components that were not included in the calculation of the DII are consumed in low amounts in Western diets such as ginger, saffron, turmeric, rosemary, flavonols, etc., and therefore, the absence of these components will likely have a negligible effect on the overall DII scores of the present population. In addition, previous published studies have also utilized this approach when calculating the DII [22,23]. Higher DII scores indicate the higher inflammatory potential of the diet.

2.3. Anthropometric Measures

Body weight was measured to the nearest 0.1 kg using a digital scale and height was measured to the nearest 0.1 cm with a wall-mounted stadiometer. Waist circumference was measured at the uppermost lateral border of the iliac crest at the narrowest point of the torso, using a flexible body tape. All measurements were done by trained research staff.

2.4. Bone Mineral Density

Bone mineral density (BMD) in g/cm^3 was measured by dual x-ray absorptiometry (DXA) in a subsample of 121 study participants. Whole body DXA scans were performed using a GE Healthcare Lunar iDXA (GE Healthcare) and analyzed with the Encore software v. 13.6, revision 2. T-scores were obtained by using peak bone mass from the manufacturer's reference population (age 30). T-scores less than -1.0 and -2.5 indicate osteopenia and osteoporosis, respectively, according to WHO criteria [24].

2.5. Other Questionnaires

A comprehensive health questionnaire was completed by study participants about demographics, physical activity (how many days per week did participant exercise for at least 20 min), lifestyle factors, reproductive health history, medication, and supplement use.

2.6. Statistical Analyses

Descriptive statistics were generated by calculating frequencies and percentages for categorical variables and means and standard deviations for continuous variables. We examined relationships between the continuous variables of interest using Pearson's correlations. Physical activity was reported in days per week of exercise and for data analysis, a dichotomous variable was created as follows: high physical activity (at least four days per week of 20 min of physical activity) and low physical activity (less than four days per week of 20 min of physical activity).

A multiple linear regression model using the backward method was fit to identify which variables were most related to WC, and the initial model included age (years), medication use category, as described below, parity (yes/no), exercise (low/high), HEI score, DII score, years since menopause, fat intake (g), and glycemic load of the diet. All variables were checked for multicollinearity prior to inclusion in the initial model. Post-hoc power analysis for the multiple regression model was calculated and the observed statistical power was 0.999.

To further explore the relationship between WC and medication use, analysis of covariance was performed including age as a covariate. Medication use was classified into three categories: low (if participants reported taking one medication or none, regularly), medium (if participants reported taking 2 to 4 medications, regularly), and high (if participants reported taking five or more medications, regularly).

To examine the relationship between diet quality and WC, linear models were fit adjusting for age and using HEI score quartile as the independent variable. We also used Pearson's or Spearman correlations and linear models to explore relationships between BMD, dietary variables, and physical activity in a subsample of participants who completed DXA scans. All p-values < 0.05 were considered statistically significant and adjustment for multiple comparisons was done by the Bonferroni test. All analyses were performed using IBM SPSS version 26.0 (IBM Corp., Armonk, NY, USA).

3. Results

3.1. Study Participants

Table 1 shows the selected characteristics of the study population. Dietary data were available for all 937 participants, but information about years since menopause and physical activity was missing for 43 and 123 participants, respectively. The majority of participants were White (97.2%) and less than 1% were Hispanic. Mean BMI was 25.1 kg/m^2 and 55.6% of the participants were classified as normal weight, while 33.5% and 10.9% were classified as overweight and obese, respectively. Between 7 and 27% of participants took a prescription medication from the categories listed in Table 1. Noticeably, while 34.7% did not report taking any prescription medications, 28.5%, 18%, and 19.5% reported taking one, two, or three or more prescription medications, respectively.

Table 1. Selected characteristics of postmenopausal women who participated in the Minnesota Green Tea Trial.

Characteristic	N	Mean (SD) or Frequency (%)	(Min–Max)
Age (years)	937	59.8 (5.0)	(50.1–71.2)
Years since menopause	894	10.7 (7.4)	(1.0–40.1)
BMI (kg/m^2)	932	25.1 (3.7)	(18.2–43.7)
Waist circumference (cm)	935	83.6 (10.2)	(60.3–120.5)
Healthy Eating Index (HEI)	937	72.6 (8.2)	(41.8–92.5)
Dietary Inflammatory Index (DII)	937	0.002 (2.2)	(−4.5–4.8)
Glycemic load	937	83.5 (32.8)	(16.4–256.9)
Protein (g)	937	58.2 (23.2)	(7.7–181.0)
Calcium from diet (mg)	937	764.5 (352.4)	(87.1–1950.3)
Calcium from supplements (mg)	937	283.2 (285.6)	(0–714.3)
Vitamin D from diet (µg)	937	3.33 (2.31)	(0.35–14.12)
Vitamin D from supplements (µg)	937	3.56 (3.32)	(0–7.14)
Physical activity	814		
Less than 4 days/week		363 (44.6)	
4 or more days/week		451 (55.4)	
Education	937		
Less than college degree		238 (25.4)	
College degree		420 (44.8)	

Table 1. *Cont.*

Characteristic	N	Mean (SD) or Frequency (%)	(Min–Max)
Graduate/Professional degree		273 (29.8)	
Parity	931		
No		218 (23.4)	
Yes		713 (76.6)	
Dietary Supplement Use	937		
No		115 (12.3)	
Yes		822 (87.7)	
Medications			
Depression/Anxiety	934	156 (16.7)	
Blood Pressure	935	187 (20.0)	
Statins	935	194 (20.7)	
Anticoagulants	935	253 (27.0)	
Thyroid	935	162 (17.3)	
Osteoporosis	935	76 (8.1)	

3.2. Dietary Intake

The mean HEI score for the participants was 72.6 (Table 1). Figure 1 depicts a radar plot of the thirteen components used to calculate the HEI stratified by WC category. The plot indicates that although the consumption of fruits, vegetables, seafood, and plant proteins, refined grains, and added sugars were within 90% of the recommended amounts, consumption of whole grains, dairy, total protein, sodium, and fats were inadequate. Consumption of whole grains and sodium met only 40% of the recommended amounts, while consumption of total protein met 70% of recommendations. In addition, the mean daily dietary calcium intake was 764.5 mg, indicating that this population was consuming only 63.7% of the recommended 1200 mg per day for women older than 51 years [16]. The intake increased to 1048 mg per day, when calcium supplements were added, which was still not sufficient to reach the dietary recommendation. Only 12.5% of participants had adequate intakes of calcium without counting with supplements. When calcium supplements were considered, 35.2% of the participants had adequate intakes, even though 68.9% reported taking a calcium supplement. Mean intakes of vitamin D from dietary sources and supplements were 3.33 µg and 3.56 µg per day, respectively. More than half of the participants reported taking a vitamin D supplement (66.6%). Mean total vitamin D was 6.88 µg, indicating that less than 50% of the recommended amount was met.

HEI score was inversely associated with DII ($r = -0.347$, $p < 0.001$) and waist circumference ($r = -0.152$, $p < 0.001$), while DII was strongly and negatively associated with glycemic load ($r = -0.676$, $p < 0.001$). DII scores ranged between -4.54 and 4.81, with a mean DII score of 0. When we looked at DII scores and glycemic load stratified by quartiles of HEI score, we found a significant association between quartiles of HEI score and DII scores, $F (3932) = 35.7$, $p < 0.001$, $\eta^2 = 0.104$. The findings indicated that mean DII scores were significantly lower for quartiles 2, 3, and 4 of HEI score, compared with quartile 1. No significant relationships were found between HEI scores and glycemic load (Table 2).

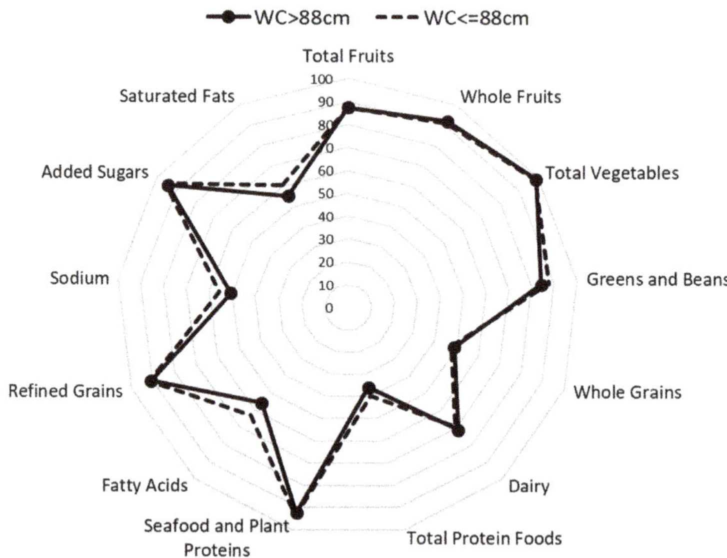

Figure 1. Radar plot showing the thirteen components of the Healthy Eating Index score stratified by WC for postmenopausal women who participated in the Minnesota Green Tea Trial. Mean HEI scores for those with WC > 88 cm and WC ≤ 88 cm were 73.7 and 75.4, respectively.

Table 2. Dietary Inflammatory Index (DII), glycemic load, and waist circumference means stratified by quartiles of HEI score ($N = 937$).

Variables	HEI Quartiles			
	Q1 (41.8–67.4)	Q2 (67.5–73.2)	Q3 (73.3–78.3)	Q4 (78.4–92.5)
DII	0.90 (0.13)	0.36 (0.13)	−0.26 (0.13)	−0.99 (0.13)
p-values [1]		0.028	0.000	0.000
Glycemic Load	86.14 (2.15)	80.63 (2.14)	80.84 (2.14)	86.27 (2.15)
p-values		NS	NS	
WC	85.89 (0.66)	84.49 (0.66)	83.39 (0.66)	80.73 (0.66)
p-values		NS	0.043	0.000

Abbreviations: DII: Dietary Inflammatory Index; HEI: Healthy Eating Index; NS: Non-Significant; WC: Waist Circumference. [1] *p*-values for comparisons with Q1.

3.3. Waist Circumference

WC measurements were available for 935 of the 937 participants. The mean of WC was 83.6 cm. When classifying participants into high and low risk of chronic disease using a WC cut-off of 88 cm [25], it was found that approximately 30% of study participants were at high risk. There was also a significant relationship between WC and HEI quartiles; after adjusting for age, F (3925) = 10.7, $p < 0.001$. WC was lower as quartiles of HEI increased, as seen in Table 2. Figure 1 shows that the components of the HEI were very similar for women classified as high risk versus those classified as low risk of chronic disease as per WC values.

We examined whether age, medication use, physical activity category, dietary (HEI score, DII score, fat intake, glycemic load), and reproductive variables (years since menopause, parity) were associated with WC by fitting a multiple linear regression model. Overall, the variables selected were not found to be good predictors of WC, as evidenced by a small $R^2 = 0.054$ ($p < 0.0001$) (Table A1). Noteworthy was the fact that mean WC for women who reported not taking any medications regularly was 82.2 cm (SE = 0.6) compared with 85.1 cm (SE = 0.6) for those who reported taking five or more medications regularly (*p* = 0.001). Women who reported exercising less than four times per week had higher mean WC (84.4 cm, SE = 0.5) compared with those who reported exercising four or more times per week (81.8 cm, SE = 0.4).

3.4. Bone Mineral Density

Mean BMD was 1.16 g/cm^3. *T*-scores ranged from -1.4 to 3.4, but only two participants were classified as osteopenic, as indicated by *T*-scores below -1.0. There were no significant correlations between BMD and HEI, total calcium, or total vitamin D intake. BMD was significantly correlated with DII (r = -0.182, *p* = 0.046) and age (r = -0.223, *p* = 0.014), and there was a trend toward a significant correlation with physical activity expressed in days per week of at least 20 min of physical activity (r$_s$ = 0.136, *p* = 0.052).

4. Discussion

The present study aimed at investigating the relationships between diet quality, assessed by calculating the HEI-2015 and the DII, and other lifestyle variables collected from a large sample of postmenopausal women who participated in the MGTT [18]. Our population's HEI score indicates that only 72.6% of the dietary guidelines were being met, with more pronounced deficits in whole grain, dairy, and total protein. Participants met recommendations for the intake of fruits and vegetables while total protein intake recommendations were not achieved, which is not in agreement with national data trends. Previously published data collected from the National Health and Nutrition Examination Survey (NHANES) in 2015–2016 showed that adults between 18 and 64 years had a HEI-2015 score of 58, with adequate intakes of total protein but noticeable deficits in fruit, vegetable, dairy, fatty acids, and whole grain intake [26]. Dairy intake in this study was 30% below the recommended amounts, which may explain the lower intake of calcium. Use of calcium supplements was reported by 68.9% of the participants and total calcium intake was improved significantly with the use of calcium supplements, increasing from 764 mg to 1048 mg (87.3% adequacy). When we looked at vitamin D intake, we found that only 45.9% of the recommended intake (15 µg/day) was met by study participants. It is thought that adequate calcium and vitamin D intake is especially important after menopause due to the decreases in bone mineral density associated with lower circulating estrogen [27], but controversy exists about the relationship between calcium and vitamin D intake and bone mineral density or risk of fractures. Bristow et al. [10] did not find any relationships between calcium intake and BMD in a cohort of osteopenic postmenopausal women over the age of 65 years. Similarly, a meta-analysis of randomized clinical trials indicated that the use of calcium and/or vitamin D supplements was not associated with a significant difference in risk of hip fractures compared to the placebo [28]. Findings from the Women's Health Initiative (WHI) randomized controlled trial of calcium plus vitamin D supplementation indicated a 1% increase in hip bone density in the group that received 1000 mg calcium with 400 IU vitamin D, but no significant effects on hip fractures were found in the intent-to-treat analysis [29]. However, in subgroup analyses, it was reported that adherent women had a 29% reduction in hip fractures. Interestingly, this trial also found that supplements of calcium plus vitamin D increased the risk of renal calculi by 17%. Jackson et al. [30] also found that hip BMD was strengthened in postmenopausal women with the combination of calcium and vitamin D supplements. We also looked at the potential relationship between calcium and vitamin D intake and BMD data that was available for 121 participants in the present dataset, but we did not find any significant associations between BMD and calcium or vitamin D intake. These findings are not entirely surprising considering that vitamin D status is influenced by several factors including sun exposure, genetics [31], and obesity [32]. Even though vitamin D intake by study participants did not meet recommendations, vitamin D status may have been adequate, which would also support the normal BMD data described here. Notwithstanding, it is recommended that postmenopausal women, particularly those living in areas where sun exposure during winter months is not sufficient to produce vitamin D in the skin, consume adequate amounts of vitamin D through the diet and/or supplements when sufficient intake cannot be achieved.

Physical activity was crudely measured in this study by asking study participants to report the number of days in which they engaged in at least 20 min of physical activity. It was found that approximately 45% of the women reported engaging in physical activity less than four times per week, while 55% engaged in four or more days per week of physical activity. In the subsample of 121 participants for whom a bone density scan was obtained, we found a nearly significant weak correlation between BMD and number of days per week of physical activity. These findings are in line with a large number of published studies examining the role of physical activity on BMD in postmenopausal women [15,33]. Unfortunately, we were not able to further explore the relationship between BMD and physical activity in the present report due to the small number of women for whom BMD data were available, and because physical activity was not assessed through a validated questionnaire or through more objective measures such as using a pedometer or an accelerometer.

DII score was calculated as another measure of diet quality in the present study. The mean DII score for our population was 0. This finding is in agreement with a previous study that reported a DII score of -0.62 for a subsample of postmenopausal women who participated in the WHI study [34]. In the WHI, DII score was significantly associated with levels of IL-6 and TNF-α-R2, which provided evidence for the validity of the DII score. Other studies have found significant associations between DII score and risk of disease such as colorectal cancer [22], cardiometabolic disease [35], and osteoporotic fractures [36]. Orchard et al. [36] used longitudinal data from the WHI to investigate the relationships between DII and risk of fractures as well as BMD measured by DXA. It was found that a higher DII score was associated with hip fracture risk only in White women younger than 63 years and the authors speculated that the much greater risk of fracture that occurs with older age would outweigh the benefits of a less inflammatory diet. The findings also indicated that lower loss of hip BMD occurred in women who consumed a less inflammatory diet (lower DII scores) over a six-year period [36]. Our findings seem to support this association between DII scores and BMD as we found a significant but weak negative correlation between BMD and DII ($r = -0.182$) in a subsample of our population. We also found a significant inverse correlation between DII score and HEI score. This was an expected finding, considering that higher DII scores indicate a more inflammatory diet, which would translate into a lower HEI. DII score was also strongly associated with glycemic load ($r = -0.676$, $p < 0.001$). To our knowledge, no previous studies have examined the associations between glycemic load and DII in postmenopausal women. In contrast with our study, Kim et al. [37] found that DII score was not correlated with glycemic load in a sample of 110 college students living in the southern U.S. These findings may be partially explained by the differences in dietary patterns between college students and postmenopausal women. For example, Kim et al. [37] reported HEI scores for female college students ranging between 34 and 61, while our HEI scores ranged from 42 to 92, indicating that postmenopausal women consume a higher quality diet.

WC and BMI were the primary anthropometric measures used in this study. Our data show that only 10.9% of study participants were classified as obese. This figure is almost four times lower than the prevalence data available for a representative sample of U.S. women, which indicated that 39.8% of non-Hispanic white women older than 20 were obese in 2017–2018 [38]. WC data indicated that 30% of participants were at high risk for disease using the cut-off value of 88 cm. Interestingly, HEI score as well as the components of the HEI were very similar between the high risk and the low risk women, indicating that differences in WC in this population may not be related to diet quality. One possible explanation for our findings could be that women who were interested in participating in the Minnesota Green Tea Trial were more likely to practice healthier lifestyle behaviors, which would lead to lower body weight. We also found that women who reported taking five or more medications regularly had higher WC compared with women who reported taking one or none, suggesting that higher WC was associated with the presence of risk factors for disease. Some of the most common medications reported by participants were anticoagulants, statins, antidepressants, blood pressure, and thyroid medications.

One limitation of the present study includes the racial background of study participants, which was primarily White and thus limits generalizability of our findings to other racial groups. It should also be noted that all women enrolled in the MGTT had been recently diagnosed with dense breasts, based on mammographic density. It is possible that these women showed higher HEI scores and lower DII scores than other populations because they were more concerned about breast cancer risk. Another limitation was the inclusion of only 28 out of the 45 food parameters used to calculate the DII score, which may have contributed to a less accurate estimation of the participants' dietary inflammatory potential. However, as described earlier, we believe that the omission of certain foods might not have significantly affected the DII scores, considering the low intake of these foods in Western cultures. The questionnaire used to assess physical activity in this population was not a validated instrument, and as such, may not have accurately assessed physical activity in this population. Finally, data on BMD was only obtained for a small subsample of 121 women, which limited the power of statistical tests that were conducted using this variable.

5. Conclusions

Our findings support the notion that White postmenopausal women living in a U.S. Midwestern state and recently diagnosed with dense breasts engage in a healthier lifestyle compared to a national sample of women older than 20 years, as evidenced by higher HEI scores and lower prevalence of obesity. In contrast, it was found that intake of dairy, calcium, and vitamin D was not adequate in this population. BMD data did not indicate the prevalence of osteopenia or osteoporosis in a subsample of the participants and, despite the lack of evidence for a strong association between calcium intake and BMD [39], future studies are needed to shed light on the relationship between calcium and vitamin D intake and the health outcomes of postmenopausal women.

Author Contributions: Conceptualization and methodology, A.Y.A. and M.S.K.; Writing—original draft preparation and data analysis, A.Y.A.; Writing—review and editing, M.S.K. All authors have read and agreed to the published version of the manuscript.

Funding: The MGTT was funded by the National Cancer Institute, grant number R01 CA127236.

Institutional Review Board Statement: The study was conducted according to the guidelines of the Declaration of Helsinki and approved by the Institutional Review Board of the University of Minnesota (protocol code 0806M36121, approved on 05/2009).

Informed Consent Statement: Written informed consent was obtained from all subjects involved in the study.

Data Availability Statement: The data presented in this study are available on request from the corresponding author. The data are not publicly available because manuscripts are still being prepared by the authors.

Conflicts of Interest: The authors declare no conflict of interest. The funders had no role in the design of the study; in the collection, analyses, or interpretation of data; in the writing of the manuscript, or in the decision to publish the results.

Appendix A

Table A1. Multiple linear regression model of predictors of waist circumference in White postmenopausal women (N = 751).

	B	SE(B)	β	*p*	95% Confidence Interval
Constant	83.974	4.953	-	0.000	74.251, 93.697
Age	0.215	0.071	0.111	0.003	0.075, 0.355
HEI score	−0.168	0.042	−0.143	0.000	−0.251, −0.085
Physical Activity [1]	−1.985	0.694	−0.102	0.004	−3.348, −0.623
Medication use [2]	−1.491	0.729	−0.075	0.041	−2.923, −0.060

B: unstandardized regression coefficient; SE(B): Standard Error of B; β: standardized coefficient. [1] Reference category is low exercise (less than four days of exercise per week). [2] Coefficient refers to category 1 (taking one or less medications regularly). Reference group is category 3 (taking five or more medications regularly).

References

1. Davis, S.R.; Lambrinoudaki, I.; Lumsden, M.; Mishra, G.D.; Pal, L.; Rees, M.; Santoro, N.; Simoncini, T. Menopause. *Nat. Rev. Dis. Primers* **2015**, *1*, 1–19. [CrossRef] [PubMed]
2. Lobo, R.A.; Davis, S.R.; De Villiers, T.J.; Gompel, A.; Henderson, V.W.; Hodis, H.N.; Lumsden, M.A.; Mack, W.J.; Shapiro, S.; Baber, R.J. Prevention of diseases after menopause. *Climacteric* **2014**, *17*, 540–556. [CrossRef] [PubMed]
3. Kapoor, E.; Collazo-Clavell, M.L.; Faubion, S.S. Weight Gain in Women at Midlife: A Concise Review of the Pathophysiology and Strategies for Management. *Mayo Clin. Proc.* **2017**, *92*, 1552–1558. [CrossRef] [PubMed]
4. Greendale, G.A.; Sternfeld, B.; Huang, M.H.; Han, W.; Karvonen-Gutierrez, C.; Ruppert, K.; Cauley, J.A.; Finkelstein, J.S.; Jiang, S.F.; Karlamangla, A.S. Changes in body composition and weight during the menopause transition. *JCI Insight* **2019**, *4*. [CrossRef]
5. Brown, L.M.; Clegg, D.J. Central effects of estradiol in the regulation of food intake, body weight, and adiposity. *J. Steroid Biochem. Mol. Biol.* **2010**, *122*, 65–73. [CrossRef] [PubMed]
6. El Khoudary, S.R.; Greendale, G.; Crawford, S.L.; Avis, N.E.; Brooks, M.M.; Thurston, R.C.; Karvonen-Gutierrez, C.; Waetjen, L.E.; Matthews, K. The menopause transition and women's health at midlife: A progress report from the Study of Women's Health across the Nation (SWAN). *Menopause* **2019**, *26*, 1213–1227.
7. Benjamin, E.J.; Muntner, P.; Alonso, A.; Bittencourt, M.S.; Callaway, C.W.; Carson, A.P.; Chamberlain, A.M.; Chang, A.R.; Cheng, S.; Das, S.R.; et al. Heart Disease and Stroke Statistics-2019 Update: A Report From the American Heart Association. *Circulation* **2019**, *139*, e56–e528. [CrossRef]
8. Finkelstein, J.S.; Brockwell, S.E.; Mehta, V.; Greendale, G.A.; Sowers, M.R.; Ettinger, B.; Lo, J.C.; Johnston, J.M.; Cauley, J.A.; Danielson, M.E.; et al. Bone mineral density changes during the menopause transition in a multiethnic cohort of women. *J. Clin. Endocrinol. Metab.* **2008**, *93*, 861–868. [CrossRef]
9. Seifert-Klauss, V.; Fillenberg, S.; Schneider, H.; Luppa, P.; Mueller, D.; Kiechle, M. Bone loss in premenopausal, perimenopausal and postmenopausal women: Results of a prospective observational study over 9 years. *Climacteric* **2012**, *15*, 433–440. [CrossRef] [PubMed]
10. Bristow, S.M.; Horne, A.M.; Gamble, G.D.; Mihov, B.; Stewart, A.; Reid, I.R. Dietary Calcium Intake and Bone Loss over 6 Years in Osteopenic Postmenopausal Women. *J. Clin. Endocrinol. Metab.* **2019**, *104*, 3576–3584. [CrossRef]
11. Tang, B.M.; Eslick, G.D.; Nowson, C.; Smith, C.; Bensoussan, A. Use of calcium or calcium in combination with vitamin D supplementation to prevent fractures and bone loss in people aged 50 years and older: A meta-analysis. *Lancet* **2007**, *370*, 657–666. [CrossRef]
12. Boonen, S.; Lips, P.; Bouillon, R.; Bischoff-Ferrari, H.A.; Vanderschueren, D.; Haentjens, P. Need for additional calcium to reduce the risk of hip fracture with vitamin D supplementation: Evidence from a comparative metaanalysis of randomized controlled trials. *J. Clin. Endocrinol. Metab.* **2007**, *92*, 1415–1423. [CrossRef]
13. Shifren, J.L.; Gass, M.L.S. The North American Menopause Society Recommendations for Clinical Care of Midlife Women. *Menopause J. N. Am. Menopause Soc.* **2014**, *21*, 1038–1062. [CrossRef]
14. Rizzoli, R.; Bischoff-Ferrari, H.; Dawson-Hughes, B.; Weaver, C. Nutrition and bone health in women after the menopause. *Womens Health* **2014**, *10*, 599–608. [CrossRef]
15. Daly, R.M.; Dalla Via, J.; Duckham, R.L.; Fraser, S.F.; Helge, E.W. Exercise for the prevention of osteoporosis in postmenopausal women: An evidence-based guide to the optimal prescription. *Braz. J. Phys. Ther.* **2019**, *23*, 170–180. [CrossRef]
16. U.S. Departments of Agriculture and Health and Human Services Dietary Guidelines for Americans, 2020–2025. Available online: https://www.dietaryguidelines.gov/resources/2020-2025-dietary-guidelines-online-materials (accessed on 9 April 2020).
17. Shivappa, N.; Steck, S.E.; Hurley, T.G.; Hussey, J.R.; Hebert, J.R. Designing and developing a literature-derived, population-based dietary inflammatory index. *Public Health Nutr.* **2014**, *17*, 1689–1696. [CrossRef]

18. Samavat, H.; Dostal, A.M.; Wang, R.; Bedell, S.; Emory, T.H.; Ursin, G.; Torkelson, C.J.; Gross, M.D.; Le, C.T.; Yu, M.C.; et al. The Minnesota Green Tea Trial (MGTT), a randomized controlled trial of the efficacy of green tea extract on biomarkers of breast cancer risk: Study rationale, design, methods, and participant characteristics. *Cancer Causes Control* **2015**, *26*, 1405–1419. [CrossRef]
19. National Institutes of Health—National Institute of Cancer—Division of Cancer Control & Population Background of the Diet History Questionnaire. Available online: https://epi.grants.cancer.gov/dhq/about/ (accessed on 9 April 2020).
20. National Institutes of Health—National Institute of Cancer—Division of Cancer Control & Population Diet History Questionnaire II: Calculating Healthy Eating INdex (HEI) Scores Using Diet*Calc Output. Available online: https://epi.grants.cancer.gov/dhq2/dietcalc/output.html (accessed on 25 February 2021).
21. Reedy, J.; Lerman, J.L.; Krebs-Smith, S.M.; Kirkpatrick, S.I.; Pannucci, T.E.; Wilson, M.M.; Subar, A.F.; Kahle, L.L.; Tooze, J.A. Evaluation of the Healthy Eating Index-2015. *J. Acad. Nutr. Diet.* **2018**, *118*, 1622–1633. [CrossRef]
22. Tabung, F.K.; Steck, S.E.; Ma, Y.; Liese, A.D.; Zhang, J.; Caan, B.; Hou, L.; Johnson, K.C.; Mossavar-Rahmani, Y.; Shivappa, N.; et al. The association between dietary inflammatory index and risk of colorectal cancer among postmenopausal women: Results from the Women's Health Initiative. *Cancer Causes Control* **2015**, *26*, 399–408. [CrossRef]
23. Duggan, C.; Tapsoba, J.d.D.; Shivappa, N.; Harris, H.R.; Hébert, J.R.; Wang, C.-Y.; McTiernan, A. Changes in Dietary Inflammatory Index Patterns with Weight Loss in Women: A Randomized Controlled Trial. *Cancer Prev. Res.* **2020**. [CrossRef]
24. World Health Organization. *Assessment of Fracture Risk and Its Application to Screening for Postmenopausal Osteoporosis: Report of a WHO Study Group [Meeting Held in Rome from 22 to 25 June 1992]*; World Health Organization: Geneva, Switzerland, 25 June 1994.
25. Janssen, I.; Katzmarzyk, P.T.; Ross, R. Body mass index, waist circumference, and health risk: Evidence in support of current National Institutes of Health guidelines. *Arch. Intern. Med.* **2002**, *162*, 2074–2079. [CrossRef]
26. Food and Nutrition Service—U.S. Department of Agriculture Healthy Eating Index. Available online: https://www.fns.usda.gov/hei-scores-americans (accessed on 17 April 2021).
27. Cosman, F.; De Beur, S.J.; Leboff, M.S.; Lewiecki, E.M.; Tanner, B.; Randall, S.; Lindsay, R. Clinician's Guide to Prevention and Treatment of Osteoporosis. *Osteoporos. Int.* **2014**, *25*, 2359–2381. [CrossRef]
28. Zhao, J.G.; Zeng, X.T.; Wang, J.; Liu, L. Association between calcium or Vitamin D supplementation and fracture incidence in community-dwelling older adults a systematic review and meta-analysis. *JAMA* **2017**, *318*, 2466–2482. [CrossRef]
29. Jackson, R.D.; LaCroix, A.Z.; Gass, M.; Wallace, R.B.; Robbins, J.; Lewis, C.E.; Bassford, T.; Beresford, S.A.A.; Black, H.R.; Blanchette, P.; et al. Calcium plus Vitamin D Supplementation and the Risk of Fractures. *N. Engl. J. Med.* **2006**, *354*, 669–683. [CrossRef] [PubMed]
30. Rossouw, J.E.; Anderson, G.L.; Prentice, R.L.; LaCroix, A.Z.; Kooperberg, C.; Stefanick, M.L.; Jackson, R.D.; Beresford, S.A.A.; Howard, B.V.; Johnson, K.C.; et al. Risks and benefits of estrogen plus progestin in healthy postmenopausal women: Principal results from the women's health initiative randomized controlled trial. *JAMA* **2002**, *288*, 321–333. [CrossRef]
31. Bahrami, A.; Sadeghnia, H.R.; Tabatabaeizadeh, S.A.; Bahrami-Taghanaki, H.; Behboodi, N.; Esmaeili, H.; Ferns, G.A.; Mobarhan, M.G.; Avan, A. Genetic and epigenetic factors influencing vitamin D status. *J. Cell. Physiol.* **2018**, *233*, 4033–4043. [CrossRef]
32. Pourshahidi, L.K. Vitamin D and obesity: Current perspectives and future directions. *Proc. Nutr. Soc.* **2015**, *74*, 115–124. [CrossRef] [PubMed]
33. Gonzalo-Encabo, P.; McNeil, J.; Boyne, D.J.; Courneya, K.S.; Friedenreich, C.M. Dose-response effects of exercise on bone mineral density and content in post-menopausal women. *Scand. J. Med. Sci. Sport.* **2019**, *29*, 1121–1129. [CrossRef]
34. Tabung, F.K.; Steck, S.E.; Zhang, J.; Ma, Y.; Liese, A.D.; Agalliu, I.; Hingle, M.; Hou, L.; Hurley, T.G.; Jiao, L.; et al. Construct Validation of the Dietary Inflammatory Index among Postmenopausal Women HHS Public Access Author manuscript. *Ann. Epidemiol.* **2015**, *25*, 398–405. [CrossRef]
35. Aslani, Z.; Sadeghi, O.; Heidari-Beni, M.; Zahedi, H.; Baygi, F.; Shivappa, N.; Hébert, J.R.; Moradi, S.; Sotoudeh, G.; Asayesh, H.; et al. Association of dietary inflammatory potential with cardiometabolic risk factors and diseases: A systematic review and dose–response meta-analysis of observational studies. *Diabetol. Metab. Syndr.* **2020**, *12*, 86. [CrossRef]
36. Orchard, T.; Yildiz, V.; Steck, S.E.; Hébert, J.R.; Ma, Y.; Cauley, J.A.; Li, W.; Mossavar-Rahmani, Y.; Johnson, K.C.; Sattari, M.; et al. Dietary Inflammatory Index, Bone Mineral Density, and Risk of Fracture in Postmenopausal Women: Results From the Women's Health Initiative. *J. Bone Miner. Res. Off. J. Am. Soc. Bone Miner. Res.* **2017**, *32*, 1136–1146. [CrossRef] [PubMed]
37. Kim, Y.; Chen, J.; Wirth, M.D.; Shivappa, N.; Hebert, J.R. Lower dietary inflammatory index scores are associated with lower glycemic index scores among college students. *Nutrients* **2018**, *10*, 182. [CrossRef] [PubMed]
38. Fryar, C.; Carroll, M.; Afful, J. Prevalence of overweight, obesity, and severe obesity among adults aged 20 and over: United States, 1960–1962 through 2017–2018. Available online: https://www.cdc.gov/nchs/data/hestat/obesity-adult-17-18/obesity-adult.htm#Citation (accessed on 18 April 2021).
39. Tai, V.; Leung, W.; Grey, A.; Reid, I.R.; Bolland, M.J. Calcium intake and bone mineral density: Systematic review and meta-analysis. *BMJ* **2015**, *351*, h4183. [CrossRef]

Review

Effects of Hormone Therapy and Flavonoids Capable on Reversal of Menopausal Immune Senescence

Nikolaos Vrachnis [1,2,3,*,†], Dimitrios Zygouris [2,†], Dionysios Vrachnis [4], Nikolaos Antonakopoulos [1], Alexandros Fotiou [5], Periklis Panagopoulos [1], Aggeliki Kolialexi [6], Kalliopi Pappa [7], George Mastorakos [8] and Zoi Iliodromiti [9]

[1] Third Department of Obstetrics and Gynecology, School of Medicine, National and Kapodistrian University of Athens, Attikon Hospital, 12462 Athens, Greece; antonakopoulos2002@yahoo.gr (N.A.); paninosrafaela@yahoo.gr (P.P.)
[2] Research Centre in Obstetrics and Gynecology, Hellenic Society of Obstetric and Gynecologic Emergency, 11528 Athens, Greece; dzygour@med.uoa.gr
[3] Vascular Biology, Molecular and Clinical Sciences Research Institute, St George's University of London, London SW17 0RE, UK
[4] Department of Clinical Therapeutics, School of Medicine, National and Kapodistrian University of Athens, Alexandra Hospital, 11528 Athens, Greece; dionisisvrachnis@gmail.com
[5] Department of Gynecologic Oncology, Metaxa Memorial Cancer Hospital, 18537 Piraeus, Greece; alexandrosfotiou92@gmail.com
[6] Department of Medical Genetics, School of Medicine, National and Kapodistrian University of Athens, 11527 Athens, Greece; akolial@med.uoa.gr
[7] First Department of Obstetrics and Gynecology, School of Medicine, National and Kapodistrian University of Athens, Alexandra Hospital, 11528 Athens, Greece; kpappa@med.uoa.gr
[8] Unit of Endocrinology, Diabetes Mellitus and Metabolism, School of Medicine, National and Kapodistrian University of Athens, Aretaieion Hospital, 11528 Athens, Greece; mastorakg@gmail.com
[9] Neonatal Department, School of Medicine, National and Kapodistrian University of Athens, Aretaieio Hospital, 11528 Athens, Greece; ziliodromiti@yahoo.gr
* Correspondence: nvrachnis@hotmail.com or nvrachnis@med.uoa.gr; Tel.: +30-697-444-1144
† These authors contributed equally to this work.

Citation: Vrachnis, N.; Zygouris, D.; Vrachnis, D.; Antonakopoulos, N.; Fotiou, A.; Panagopoulos, P.; Kolialexi, A.; Pappa, K.; Mastorakos, G.; Iliodromiti, Z. Effects of Hormone Therapy and Flavonoids Capable on Reversal of Menopausal Immune Senescence. *Nutrients* **2021**, *13*, 2363. https://doi.org/10.3390/nu13072363

Academic Editors: Emad Al-Dujaili and Giuseppina Mandalari

Received: 28 April 2021
Accepted: 7 July 2021
Published: 10 July 2021

Publisher's Note: MDPI stays neutral with regard to jurisdictional claims in published maps and institutional affiliations.

Copyright: © 2021 by the authors. Licensee MDPI, Basel, Switzerland. This article is an open access article distributed under the terms and conditions of the Creative Commons Attribution (CC BY) license (https://creativecommons.org/licenses/by/4.0/).

Abstract: Menopause, probably the most important natural change in a woman's life and a major component of female senescence, is characterized, inter alia, by cessation of ovarian estrogen and progesterone production, resulting in a gradual deterioration of the female immune system. Hormone replacement therapy (HRT) is used in postmenopausal women to relieve some of the peri- and postmenopausal symptoms, while there is also evidence that the therapy may additionally partially reverse menopausal immune senescence. Flavonoids, and especially isoflavones, are widely used for the treatment of menopausal symptoms, although it is not at present clear whether they can reverse or alleviate other menopausal changes. HRT reverses the menopausal CD4/CD8 ratio and also limits the general peri- and postmenopausal inflammatory state. Moreover, the increased levels of interleukins (IL)-1β, IL-6, and IL-8, as well as of tumor necrosis factor-α (TNF-α) are decreased after the initiation of HRT. However, some reports show no effect of HRT on IL-4, IL-10, and IL-12. It is thus evident that the molecular pathways connecting HRT and female immune senescence need to be clarified. Interestingly, recent studies have suggested that the anti-inflammatory properties of isoflavones possibly interact with inflammatory cytokines when applied in menopause treatments, thereby potentially reversing immune senescence. This narrative review presents the latest data on the effect of menopausal therapies, including administration of flavonoid-rich products, on age-associated immune senescence reversal with the aim of revealing possible directions for future research and treatment development.

Keywords: flavonoids; isoflavones; hormone replacement therapy; menopause; immune senescence; inflammation; inflammatory cytokines; immune system

1. Introduction

Menopause, which is characterized by the termination of the ovarian production of estrogen and progesterone and is considered to be the most important natural change in a woman's life, is accompanied by a large number of correlated changes. In order to alleviate some of these, such as menopausal vasomotor symptoms and genital atrophy, and to prevent or reduce bone loss, hormone replacement therapy (HRT) is frequently used by peri- and postmenopausal women. However, there are a number of additional applications of hormone treatment (HT), in which hormones are administered for other medical purposes in humans or tested in animal models.

Flavonoids are a class of polyphenolic compounds: isoflavones are the most widespread, mainly found in legumes, grains, nuts, and vegetables, with soybeans being the richest source. They have been used successfully for the treatment of menopausal symptoms and have also shown a favorable effect on prevention and treatment of osteoporosis in postmenopausal women [1–3].

Another result of the menopausal transition is the senescence, or aging, of the female immune system. In this setting, a reduced immune response develops, as well as an increased inflammatory state, which condition results in the triggering or exacerbation of an array of disorders [4].

The last decade has seen the publication of several studies reporting the considerable beneficial effect exerted by female sex hormones both on postmenopausal conditions and on the immune system, especially with regard to autoimmune diseases and infections [5]. The authors, meanwhile, also demonstrated the manner in which low estrogen levels compromise the immune response, thus predisposing women to disease and infection [6–9]. A number of studies have, moreover, pointed to the contribution of ovarian sex steroid loss towards immune senescence among women, while also showing that menopausal hormone therapy (MHT) is capable of delaying several of these changes [10]. In addition, recent in vitro studies have revealed the anti-inflammatory effects of isoflavones, mentioned above, as evidenced by a reduction in the levels of inflammatory cytokines [11–13].

Since it is well-known that ovarian steroids modulate the immune response, it could be suggested that HRT may possibly reverse immune senescence. Nevertheless, the exact molecular pathways connecting HRT and immune senescence have not to date been elucidated. Chronic inflammation, promoted by certain lifestyle factors, is known to play a key role in human health by inducing diseases that lead to morbidity and mortality. It is of interest to note that the increased vascular inflammation [14] observed in women in early menopause, as compared to age-matched premenopausal women, is mirrored by a similar phenomenon that manifests from birth in premature newborns of both sexes [15,16], a condition which lasts throughout life [17–20].

Given that reduction of inflammation shows clear improvement in certain cases, deeper insight into the molecular pathways of the inflammatory cascade may well offer highly promising options for future treatments [21].

In this narrative review, we present the most recent data concerning the impact of hormonal menopausal therapies and isoflavones on the immune system and potential, resultant senescence reversal with the aim of revealing possible directions for future research and treatment development, while we also discuss flavonoid-rich products that may be used for pharmaceutical purposes or as functional foods.

2. Basic Functions of the Immune System Related to Immune Senescence

The prime function of the human immune system is to provide protection from external pathogens and to fight infection: this is accomplished via its recognition of self and non-self antigens, while it also assesses microbial threats and deals with their elimination [22]. It is, moreover, essential for minimizing any damage done to tissues, an action that inhibits the development of such disorders as autoimmune diseases. This extraordinary process of fine-tuning is made possible via the subtle interplay between the two vital strands of the immune system, namely, innate and adaptive immunity [23].

The actions of the innate immune system, which constitutes the body's first line of defense against diseases nonspecific for any pathogen, are mediated by natural killer (NK) cells, neutrophils, macrophages, and dentritic cells. All the latter identify microbial nonself pathogens by means of pattern recognition repeaters (PRRS), of which toll-like receptors (TLR) compose the best characterized family. NK cells mediate recognition of missing or altered self through expression of an array of activating and inhibitory receptors. Macrophages and neutrophils, meanwhile, participate in the removal of pathogens via phagocytosis, whereas activation of these macrophages and neutrophils results in an inflammatory response on account of the formation of interleukin (IL)-6, IL-8, and interferon-α (IFN-α). Finally, IL-6, acting as a proinflammatory cytokine, activates IL-1 and IL-10 while also inhibiting tumor necrosis factor-α (TNF-α): it thereby stimulates the autoimmune and inflammatory process in a range of diseases. IL-8, a neutrophil chemotactic factor, is increased by oxidant stress, inducing localized inflammation by cleaving metalloproteinase molecules [24]. IFN-α, an inflammatory cytokine that may be inhibited by IL-10, induces immune dysfunction in autoimmune diseases while also mediating tissue inflammation [25].

The two types of lymphocytes of the adaptive immune response are T and B cells, which are able to recognize and respond specifically to each pathogen. B cells, or B lymphocytes, produce large quantities of antibodies specific to an antigen, while T lymphocytes, which are categorized into one cluster of differentiation 4 (CD4) cells and one cluster of differentiation 8 (CD8) cells, identify small peptides as antigens, while inflammatory molecules, such as IL-7 and IL-5, play an essential part in T-cell homeostasis [26]. CD4, which acts as a co-receptor of the T-cell receptor (TCR), assists the latter through its communication with antigen-presenting cells, while both CD4 and the TCR complex bind to separate regions of the antigen-presenting MHC class II molecule, thereby helping to stabilize weak TCR interactions and enhancing TCR signaling [27].

Finally, it is important to underline that COVID-19 infection is characterized by a significantly better outcome in women, with male sex being considered a poor prognostic factor. A possible explanation for the latter may be the interaction between sex hormone levels and the immune system in women during reproductive age [28].

3. Hormone Treatment and the Immune System

Over the last two decades, numerous studies have been conducted to investigate the impact of hormone treatment on the immune system. The interactions between estrogen, progesterone, HRT, and the cells of the immune system as observed in women and in animal models are explored below.

3.1. Estrogen and the Human Immune System

A very interesting finding is that high estrogen levels in humans are associated with better vaginal immunization [29]. In vitro studies in humans similarly found that the presence of E2 promoted maturation of dendritic cells and differentiation of naïve CD4 T cells into T helper type 2 (TH2) cells [30]. On the other hand, it is reported that estradiol (E2) limits the efficiency of mature dendritic cells and their ability to stimulate T-cell proliferation [31].

Estrogen administration promotes production of Bcl-2, an anti-apoptotic molecule in B cells, and it is suggested that it subsequently increases the autoreactive resistance of B cells to apoptosis [32].

A recent study, moreover, demonstrated that estrogens lead to greater antibody affinity maturation by upregulating activation-induced deaminase (AID) and increasing the somatic hypermutation frequency and class-switch recombination [33].

3.2. Estrogen Treatment in Animals: Effects on Immune System

A large number of experimental data show that estrogens promote the protective Th1 response that follows parasitic infections. More specifically, in an animal model of parasitic infections, hormone treatment with 17-β estradiol (E2) in male mice resulted in higher expression of IFN-γ and lower levels of IL-10 in spleen cells [34]. The same effect was also noted in female ovariectomized mice treated with E2, which induced production of significantly higher levels of IFN-γ and antibodies [35], while HRT treatment in this model caused a lower degree of weight loss and of hematocrit diminution.

The protective role of estrogens has additionally been demonstrated in another animal study in which E2 treatment increased resistance to Toxoplasma gondii infections in female and also in male mice models [36].

In a mouse model, E2 administration to ovariectomized females resulted in a protective effect against HSV-2 infection [37]. The same protection was also observed in female rhesus macaques against transmission of SIV (simian immunodeficiency virus): this is a retrovirus that causes persistent infections in 45 or more species of African nonhuman primates [38]. In another murine HSV-2 challenge model, a study was made of the immune response after vaccination of ovariectomized females treated with E2. Ovariectomized females exhibited the same immune response to vaccination as unvaccinated controls, while the administration of E2 enhanced protection and significantly decreased disease severity [39]. However, the antibody titers in the E2-treated group were not found to be significantly higher compared to those of the control group, with only improved neutralization being reported.

It has, in addition, been determined that the estrogen-related receptor (ERR) effect on macrophages is necessary for the adequate production of interferon gamma (IFN-γ) and for an efficient immune response after infection with Listeria monocytogenes [40]. It is also observed that estrogens enhance B-cell activation and IgG production [41].

Concerning the effect of estrogens on T cells by use of peripheral blood mononuclear cells (PBMC), until now the reported data remain contradictory [42]. However, it is of note that in vitro studies point to a potential bias towards a Th2 response [43] and regulatory T (Treg) polarization [44] following E2 administration.

3.3. Progestogen and the Immune System in Humans and Animals

An association between Depo-Provera and increased incidence of chlamydia, HSV-2, HI, and HPV was confirmed with clinical studies in adult women [45]. On the other hand, progesterone is found to promote T-cell response [40], while it simultaneously diminishes the efficiency of dendritic cells (DCs) in the uptake of antigenic peptides [46]. Macrophages and DCs that were exposed to progesterone revealed a lower state of activation compared to that of untreated cells [47] and reduced apoptosis in NK cells [48]. Of note, progesterone is able to induce skewing of CD4 T-cell responses to Th2-type responses [49].

The same phenomenon was found in female animal models with prolonged exposure to progesterone, which resulted in increased susceptibility to HSV-2 genital infection [50], chlamydia trachomatis [51], SIV [52], and SHIV [53]. Moreover, progesterone treatment in mouse models reduced production of nitric oxide (NO) and macrophage activation [54].

3.4. HRT and the Immune System

In human studies, it was found that women taking HRT have higher lymphocyte numbers compared with post-menopausal women not receiving HRT [55]. This difference is already obvious within the first month after treatment initiation. There are also indications that HRT restores the levels of circulating monocytes in menopausal women to levels seen in women with menstruation [56].

HRT specifically counteracts the effects of menopause on B cells. It is hypothesized that the menopause brings about a large reduction in B cells, and specifically in B2 cell production; after administration of HRT B cells, production is increased [57]. By contrast, other human studies have demonstrated no effect of HRT on the menopausal changes to the T-cell line [55,58,59]. More explicitly, no improvement in the women's health was observed with the reduction in the percentage of naïve T cells, the elevation of activated T cells, or the accumulation of memory T cells. However, this could possibly be attributed to the subjects' biological aging rather than to menopausal transition. Also to be considered are the different types of hormonal therapy used in the various studies, which may, while relieving menopausal symptoms, also, to varying degrees, slow down the menopausal changes of the immune system.

The CD4/CD8 ratio of T cells decreases with age and is associated with reduced resistance to infection, while the percentage of NK cells is increased compared to that in young women. It is well-known that after menopause, this ratio is significantly reduced, together with the actual number of circulating B cells. HRT has been found to substantially neutralize these changes, while there are studies reporting that combined hormone therapy resulted in immediate increase of circulating B cells and especially in B2 population of B cells, as well as of T cells to a lesser extent [60,61]. Of special note, while estrogen therapy in hysterectomized women did not bring about a significant change in CD4 T cells or NK cells [62], in contrast, another study in menopausal nonhysterectomized women [63] following estrogen therapy demonstrated a reduction in CD8 T cells and a subsequent rise in the CD4/CD8 ratio, this revealing improved status of the immune system.

4. HRT and Specific Inflammatory Molecules

4.1. Human Studies

Menopause is in general characterized as an inflammatory state, which is mainly due to the increased levels of proinflammatory and inflammatory cytokines [2]. Various studies investigating the effect of HRT on inflammatory molecules have, however, differed significantly with regard to the regimen used for HRT (transdermal estrogen, tibolone), the daily dosage, and the duration of administration (3 to 6 months). Most of the data come from studies including mainly women in early postmenopause, a time period during which there is a continuous increase of FSH and decrease of estradiol—this lasting for about 2 years following the last menstrual period (STRAW+10 criteria) [64].

It has been established that in postmenopausal women, IL-1β, TNF-α, and IL-6 are significantly higher compared to levels in women of reproductive age [65]. HRT also increases the expression of IL-6 and soluble Il-6 receptor [66]; however, the latter menopausal inflammatory state characterized by increased cytokine concentrations is not detected or is noted to a lesser degree in women on estrogen therapy [67]. While the existing studies on inflammatory cytokines remain controversial, they nonetheless generally point to a decline in menopausal inflammation after HRT (Table 1; Table 2).

More specifically, IL-6 levels are seen to have decreased in postmenopausal women receiving transdermal estrogen as soon as 3 months after the first prescription [68]. Therefore, it is suggested that IL-6 levels have a negative correlation with transdermal estrogens in HRT users, which is the same as that seen in women in menopause aged 40 to 65 years without any treatment. In another, smaller, human study though, IL-6 levels were found to be higher in women taking combined HRT compared to nonusers after PBMC stimulation by lipopolysaccharide (LPS) [67].

There are data showing no reversal of TNF-α increase in women who are on HRT [69], while levels of IL-6 are negatively correlated with serum estradiol concentration [70], leading to reduced inflammatory response with increased estradiol.

Table 1. Inflammatory molecule actions in menopausal women and after HRT use for menopause (NS: not studied).

Mediator	Molecular Weight (Da)	Menopause (Humans)	HRT (Hormone Replacement Treatment)	HRT Regimen	Dose	Duration	Study Participants
IL-1β	30,748	Increased [1]	No effect [70]	Low transdermal estrogen	1 mg/day	3 months	66
IL-4	17,492	Controversial data [1,67]	No effect [67]	Tibolone	2.5 mg/day	6 months	49
IL-6	23,718	Increased [1,62,68]	Decreased [68]	Low-dose transdermal estrogen	1 mg/day	3 months	66
IL-8	11,098	Increased [1,68]	NS				
IL-10	20,517	Increased [1,67]	No effect [66,67]	Transdermal estrogen	50 microg/day	3 months	30
				Tibolone	2.5 mg/day	6 months	49
IL-12	24,874	Increased [1,67]	No effect [67]	Tibolone	2.5 mg/day	6 months	49
TNF-α	25,644	Increased [1,62,67]	NS				
			No effect [67,70]	Low-dose transdermal estrogen	1 mg/day	3 months	66
				Tibolone	2.5 mg/day	6 months	49

The changes of IL-10 and IL-12 levels in menopausal transition are also not as yet fully clarified, with some studies reporting an increase and others a decrease during menopause [4]. However, HRT and transdermal estrogen have shown no effect on IL-10 levels [68,69]. With regard to IL-8, there is a negative correlation with estrogen levels in both human [70] and animal studies [71].

Another study demonstrated elevated markers of vascular repair along with improved microvascular reactivity following short-term administration of low-dose transdermal estradiol therapy in overweight/obese menopausal women, though no change was noted in levels of inflammatory molecules, including IL-1β, IL-6, MCP-1, and TNF-α [72].

IFN-γ is observed to be decreased in menopausal women and in women who have undergone bilateral oophorectomy [4]. After the initial treatment, IFN-γ increases [62]—while combined hormone therapy has been found to reduce IFN-γ production [73], this possibly attributable to progesterone's opposing effect.

Tibolone, a synthetic steroid with estrogenic, progestogenic, and androgenic properties, is used widely as a treatment option in menopausal women. To date, there are, however, limited data concerning its effect on the immune system. In a human study, it was reported that tibolone had no effect on reversing the menopausal impact on IL-4, IL-10, IL-12, and TNF-α plasma levels [69], probably due to a combination of the above properties.

4.2. Animal Studies

In Table 2, we present all the data from animal studies that used primates, as well as menopause study models with other animals. It must be borne in mind that most animal models were rats, with different life expectancy and with mostly gonadectomy-induced menopause.

It has been noted that chronic E2 administration in ovariectomized animal models considerably increases the inflammatory cytokines IL-1β, IL-6, and TNF-α, and also of NO synthase in thioglycate-elicited macrophages [74], confirming previous reports that showed anti-inflammatory effects from short-term exposure to estrogens and enhanced production of proinflammatory cytokines in chronic exposure [75,76]. Furthermore, a recent study in ovariectomized rats reported that serum levels of TNF-α and IL-6 were increased, while treatment with estrogen or raloxifen or tamoxifen normalized these levels [77].

A study in ovariectomized rats showed that the increased levels of IL-6 and TNF-α were reduced after treatment with 17-β estradiol [78]. The same study reported that treatment with tibolone additionally reduced TNF-α, while there was no impact on IL-6 levels [78]. The same effect was also demonstrated in atrial natriuretic peptide (ANP) concentration, which was increased after HRT with tibolone, implying that tibolone has better anti-inflammatory effects compared to those of 17β-estradiol and could prevent cardiovascular disease.

Table 2. Inflammatory molecule actions in animal studies in menopause and after HRT use in menopause (NS: not studied).

Mediator	Menopause (Humans)	HRT (Animals)	HRT Regimen
IL-1β	Increased [1]	Decreased [75]	Estrogen, raloxifen, tamoxifen
IL-4	Controversial data [1,67]	NS	
IL-6	Increased [1,62,68]	Decreased [78]	Estrogen, raloxifen, tamoxifen
		No effect [78]	Tibolone
IL-8	Increased [1,68]	NS	
IL-10	Increased [1,67]	NS	
IL-12	Increased [1,67]	NS	
TNF-α	Increased [1,62,67]	Decreased [77,78]	Estrogen, raloxifen, tamoxifen, tibolone

5. Nutritional Supplementation with Flavonoids and the Immune System

Apart from hormone specimens, alternative approaches are additionally currently being developed for the alleviation of menopausal symptoms. Among them, nutritional supplementation with estrogen-like substances has, over the last decade, attracted a considerable amount of research interest.

Flavonoids are a class of polyphenolic secondary metabolites that are found in plants and are regularly consumed in the human diet. Until now more than 5000 flavonoids from various plants have been characterized. Based on their chemical structure, they are divided into subgroups, with the subgroup of isoflavones being the most commonly found, and are particularly important since they act as phytoestrogens in humans [79]. Isoflavones are mainly contained in soybeans, green beans, mung beans, cowpeas, kudzu roots, red clover sprouts, soy milk, peanuts (genistein), and green tea. The richest source of isoflavones is found in soybeans, with soy-derived foods and ingredients containing various concentrations of isoflavones.

In a human study of menopausal women taking either soy milk with 71.6 mg isoflavones or a supplement of 70 mg isoflavones for 16 weeks, an increase in circulating B cells was detected [80], while in an in vitro model, isoflavones resulted in increased IL-10 production [81]. The same trend of enhancement of the activities of cytotoxic T cells and NK cells was also observed in a human model using daidzein and genistein [82] and in a murine model [83] investigating the isoflavone genistein. Another animal study of ovariectomized female rats reported that nutritional supplementation with soybean or soybean and green tea enhanced the proliferative potential of B and T cells as well as of NK cell killing [84]. The same study, moreover, noted improved chemotaxis and macrophage production of reactive oxygen species (ROS), which have been demonstrated as being associated with ovarian function and estrogen levels [85].

Moreover, soymilk with 71.6 mg isoflavones or a supplement of 70 mg isoflavones for 16 weeks have been found to result in higher B-cell populations in healthy postmenopausal women, while the compounds have, in addition, been shown to have a protective role against DNA damage in menopausal women [80].

6. Isoflavones and Inflammation in Menopausal Women

There are data revealing a reduction in the inflammatory molecules IL-6 and TNF-α in women's plasma levels after they received nutritional supplementation with isoflavones [86–88] (Table 3).

Chi et al., using soy isoflavones at a dose of 90 mg/day for 6 months, noted a reduction in the levels of IL-6 and TNF-α [86]. Moreover, Nadadur et al. also observed a reduction in levels of TNF-α, but no effect on IL-6 levels using food supplementation with either 50 mg isoflavones or 15 g soy protein in the form of tofu for 8 weeks [87]. Another study with 80 mg isoflavones (60.8 mg of genistein, 16 mg of daidzein, and 3.2 mg of glicitein) for 6 months showed reductions in TNF-α levels [88].

Recent in vitro studies have described the anti-inflammatory effects of isoflavones on peripheral blood mononuclear cells stimulated with lipopolysaccharide, thereby reducing IL1-β, IL-2, IL-6, and TNF-α levels [89,90].

Table 3. Inflammatory molecule actions in menopausal women and in vitro experiments after isoflavones use (NS: not studied).

Mediator	Molecular Weight (Da)	Menopause (Humans)	Isoflavones (Humans)	Isoflavones (In Vitro)
IL-1β	30,748	Increased [1]	No effect [91]	Decreased [89,90]
IL-4	17,492	Controversial data [1,67]	NS	NS
IL-6	23,718	Increased [1,52,68]	Decreased [86] / No effect [87,91,92]	Decreased [89]
IL-8	11,098	Increased [1,68]	No effect [93]	NS
IL-10	20,517	Increased [1,67]	NS	NS
IL-12	24,874	Increased [1,67]	NS	NS
TNF-α	25,644	Increased [1,62,67]	Decreased [86–88] / No effect [80,91]	Decreased [89,90]

Nevertheless, it must also be stressed that studies have been published showing no impact of this alternative approach on the immune systems of menopausal women. Moreover, a number of studies demonstrate no change in IL-2, TNF-α, and IFN-γ levels [80,91]. Among the latter are the study of Ryan-Borchers et al., who used either soy milk with 71.6 mg isoflavones or a supplement of 70 mg isoflavones for 16 weeks [80], and that of Beavers et al., who conducted a trial with soy milk for 4 weeks [91]. Other reports demonstrated no change in IL-6 levels with either 50 mg isoflavones or 15 g soy protein in the form of tofu for 8 weeks [87]; nor did the study by Huang et al., using soy milk with 112 mg isoflavones for 8 weeks, show any positive results [92]. Similarly concerning IL-8 levels, no change was detected after a supplement intake of 100 mg isoflavones (33 mg of genistein, 93.5 mg of dadzein, and 3.2 mg of glycitein) for 10 weeks [93]. The above data point to the fact that there is as yet insufficient evidence to support nutritional supplementation with isoflavones for enhancement of the immune response among menopausal women.

7. Conclusions

Taken together, the above-reported data reveal that the impacts of HRT on the immune system and its senescence remain to date unclear in many respects. Nevertheless, what appears evident is that HRT reverses the menopausal CD4/CD8 ratio and moderates the overall inflammatory state. Another beneficial effect is the fact that the increased levels of interleukins 1β, 6, 8, and TNF-α are reversed after HRT. As concerns other molecules, no effect of HRT has been observed on IL-4, 10, and 12.

The discrepancies noted between the available studies are partially attributable both to the large variety of current hormone replacement regimens, as these are shown in Table 1, and to the lack of consensus regarding the optimum immune parameters to be applied. It is thus necessary for more studies, both experimental and clinical, to be conducted in order to elucidate the observed alterations in concentrations in other mediators of the molecular pathways connecting HRT and the immune system.

An exact understanding of the impact of HRT and isoflavones on the functioning of the immune system will be the cornerstone for establishing targeted therapies for menopausal women. These future therapies will have the advantage of overcoming the effects of menopause without the side effects of the existing treatment options. At the same time, flavonoid-rich products, such as flavonoid-rich green tea, isoflavone-rich soy, flax seed, flavonol quercetin, and isoflavones, could be the basis for the development of pharmaceutical products and/or functional foods in the future.

Another interesting aspect is that targeted therapies may also improve the immune response to infection, inflammation, and vaccination of menopausal women, leading to the subsequent reduction of the consequent morbidity and mortality in this age group.

To conclude, an important focus of future research should be assessment of disease severity and mortality in menopausal women who have received either hormone therapy or flavonoids, compared to women who have had no treatment, or men of the same age group.

Author Contributions: Conceptualization, N.V., D.Z., and Z.I.; methodology, D.Z., D.V., and N.A.; writing—original draft preparation, N.V., D.Z., D.V., N.A., A.F., and N.A.; writing—review and editing, N.V., D.Z., P.P., A.K., K.P., G.M., and Z.I. All authors have read and agreed to the published version of the manuscript.

Funding: This research received no external funding.

Conflicts of Interest: The authors declare no conflict of interest.

References

1. Gomez-Zorita, S.; Gonzalez-Arceo, M.; Fernandez-Quintela, A.; Eseberri, I.; Trepiana, J.; Portillo, M.P. Scientific Evidence Supporting the Beneficial Effects of Isoflavones on Human Health. *Nutrients* **2020**, *12*, 3853. [CrossRef] [PubMed]
2. Abdi, F.; Rahnemaei, F.A.; Roozbeh, N.; Pakzad, R. Impact of phytoestrogens on treatment of urogenital menopause symptoms: A systematic review of randomized clinical trials. *Eur. J. Obstet. Gynecol. Reprod. Biol.* **2021**, *261*, 222–235. [CrossRef] [PubMed]
3. Sansai, K.; Na Takuathung, M.; Khatsri, R.; Teekachunhatean, S.; Hanprasertpong, N.; Koonrungsesomboon, N. Effects of isoflavone interventions on bone mineral density in postmenopausal women: A systematic review and meta-analysis of randomized controlled trials. *Osteoporos. Int.* **2020**, *31*, 1853–1864. [CrossRef] [PubMed]
4. Vrachnis, N.; Zygouris, D.; Iliodromiti, Z.; Daniilidis, A.; Valsamakis, G.; Kalantaridou, S. Probing the impact of sex steroids and menopause-related sex steroid deprivation on modulation of immune senescence. *Maturitas* **2014**, *78*, 174–178. [CrossRef]
5. Desai, M.K.; Brinton, R.D. Autoimmune Disease in Women: Endocrine Transition and Risk Across the Lifespan. *Front. Endocrinol.* **2019**, *10*, 265. [CrossRef]
6. Kadel, S.; Kovats, S. Sex Hormones Regulate Innate Immune Cells and Promote Sex Differences in Respiratory Virus Infection. *Front. Immunol.* **2018**, *9*, 1653. [CrossRef]
7. Kovats, S. Estrogen receptors regulate innate immune cells and signaling pathways. *Cell. Immunol.* **2015**, *294*, 63–69. [CrossRef]
8. Giefing-Kroll, C.; Berger, P.; Lepperdinger, G.; Grubeck-Loebenstein, B. How sex and age affect immune responses, susceptibility to infections, and response to vaccination. *Aging Cell* **2015**, *14*, 309–321. [CrossRef]
9. Angelidis, G.; Dafopoulos, K.; Messini, C.I.; Valotassiou, V.; Tsikouras, P.; Vrachnis, N.; Psimadas, D.; Georgoulias, P.; Messinis, I.E. The emerging roles of adiponectin in female reproductive system-associated disorders and pregnancy. *Reprod. Sci.* **2013**, *20*, 872–881. [CrossRef]
10. Abdi, F.; Mobedi, H.; Mosaffa, N.; Dolatian, M.; Ramezani Tehrani, F. Effects of hormone replacement therapy on immunological factors in the postmenopausal period. *Climacteric.* **2016**, *19*, 234–239. [CrossRef]
11. Maleki, S.J.; Crespo, J.F.; Cabanillas, B. Anti-inflammatory effects of flavonoids. *Food Chem.* **2019**, *299*, 125124. [CrossRef]
12. Asbaghi, O.; Yaghubi, E.; Nazarian, B.; Kelishadi, M.R.; Khadem, H.; Moodi, V.; Naeini, F.; Ghaedi, E. The effects of soy supplementation on inflammatory biomarkers: A systematic review and meta-analysis of randomized controlled trials. *Cytokine* **2020**, *136*, 155282. [CrossRef]
13. Khodarahmi, M.; Foroumandi, E.; Asghari Jafarabadi, M. Effects of soy intake on circulating levels of TNF-alpha and interleukin-6: A systematic review and meta-analysis of randomized controlled trials. *Eur. J. Nutr.* **2021**, *60*, 581–601. [CrossRef]

14. Bechlioulis, A.; Naka, K.K.; Kalantaridou, S.N.; Kaponis, A.; Papanikolaou, O.; Vezyraki, P.; Kolettis, T.M.; Vlahos, A.P.; Gartzonika, K.; Mavridis, A.; et al. Increased vascular inflammation in early menopausal women is associated with hot flush severity. *J. Clin. Endocrinol. Metab.* **2012**, *97*, E760–E764. [CrossRef]

15. Vrachnis, N.; Karavolos, S.; Iliodromiti, Z.; Sifakis, S.; Siristatidis, C.; Mastorakos, G.; Creatsas, G. Review: Impact of mediators present in amniotic fluid on preterm labour. *In Vivo* **2012**, *26*, 799–812.

16. Vrachnis, N.; Vitoratos, N.; Iliodromiti, Z.; Sifakis, S.; Deligeoroglou, E.; Creatsas, G. Intrauterine inflammation and preterm delivery. *Ann. N. Y. Acad. Sci.* **2010**, *1205*, 118–122. [CrossRef]

17. Iliodromiti, Z.; Antonakopoulos, N.; Sifakis, S.; Tsikouras, P.; Daniilidis, A.; Dafopoulos, K.; Botsis, D.; Vrachnis, N. Endocrine, paracrine, and autocrine placental mediators in labor. *Hormones* **2012**, *11*, 397–409. [CrossRef]

18. Vrachnis, N.; Malamas, F.M.; Sifakis, S.; Tsikouras, P.; Iliodromiti, Z. Immune aspects and myometrial actions of progesterone and CRH in labor. *Clin. Dev. Immunol.* **2012**, *2012*, 937618. [CrossRef]

19. Deligeoroglou, E.; Vrachnis, N.; Athanasopoulos, N.; Iliodromiti, Z.; Sifakis, S.; Iliodromiti, S.; Siristatidis, C.; Creatsas, G. Mediators of chronic inflammation in polycystic ovarian syndrome. *Gynecol. Endocrinol.* **2012**, *28*, 974–978. [CrossRef]

20. Vrachnis, N.; Belitsos, P.; Sifakis, S.; Dafopoulos, K.; Siristatidis, C.; Pappa, K.I.; Iliodromiti, Z. Role of adipokines and other inflammatory mediators in gestational diabetes mellitus and previous gestational diabetes mellitus. *Int. J. Endocrinol.* **2012**, *2012*, 549748. [CrossRef]

21. Iliodromiti, Z.; Zygouris, D.; Sifakis, S.; Pappa, K.I.; Tsikouras, P.; Salakos, N.; Daniilidis, A.; Siristatidis, C.; Vrachnis, N. Acute lung injury in preterm fetuses and neonates: Mechanisms and molecular pathways. *J. Matern. Fetal Neonatal Med.* **2013**, *26*, 1696–1704. [CrossRef] [PubMed]

22. Goldman, A.S.; Prabhakar, B.S. Immunology Overview. In *Medical Microbiology*; Th Baron, S., Ed.; The University of Texas Medical Branch at Galveston: Galveston, TX, USA, 1996.

23. Justiz Vaillant, A.A.; Jan, A. *Physiology, Immune Response*; StatPearls: Treasure Island, FL, USA, 2020.

24. Barouchos, N.; Papazafiropoulou, A.; Iacovidou, N.; Vrachnis, N.; Barouchos, N.; Armeniakou, E.; Dionyssopoulou, V.; Mathioudakis, A.G.; Christopoulou, E.; Koltsida, S.; et al. Comparison of tumor markers and inflammatory biomarkers in chronic obstructive pulmonary disease (COPD) exacerbations. *Scand. J. Clin. Lab. Investig.* **2015**, *75*, 126–132. [CrossRef] [PubMed]

25. Nadkarni, S.; McArthur, S. Oestrogen and immunomodulation: New mechanisms that impact on peripheral and central immunity. *Curr. Opin. Pharmacol.* **2013**, *13*, 576–581. [CrossRef] [PubMed]

26. Sakiani, S.; Olsen, N.J.; Kovacs, W.J. Gonadal steroids and humoral immunity. *Nat. Rev. Endocrinol.* **2013**, *9*, 56–62. [CrossRef]

27. Glatzova, D.; Cebecauer, M. Dual Role of CD4 in Peripheral T Lymphocytes. *Front. Immunol.* **2019**, *10*, 618. [CrossRef]

28. Zheng, Z.; Peng, F.; Xu, B.; Zhao, J.; Liu, H.; Peng, J.; Li, Q.; Jiang, C.; Zhou, Y.; Liu, S.; et al. Risk factors of critical & mortal COVID-19 cases: A systematic literature review and meta-analysis. *J. Infect.* **2020**, *81*, e16–e25.

29. Kozlowski, P.A.; Williams, S.B.; Lynch, R.M.; Flanigan, T.P.; Patterson, R.R.; Cu-Uvin, S.; Neutra, M.R. Differential induction of mucosal and systemic antibody responses in women after nasal, rectal, or vaginal immunization: Influence of the menstrual cycle. *J. Immunol.* **2002**, *169*, 566–574. [CrossRef]

30. Uemura, Y.; Liu, T.Y.; Narita, Y.; Suzuki, M.; Matsushita, S. 17 Beta-estradiol (E2) plus tumor necrosis factor-alpha induces a distorted maturation of human monocyte-derived dendritic cells and promotes their capacity to initiate T-helper 2 responses. *Hum. Immunol.* **2008**, *69*, 149–157. [CrossRef]

31. Segerer, S.E.; Muller, N.; van den Brandt, J.; Kapp, M.; Dietl, J.; Reichardt, H.M.; Rieger, L.; Kammerer, U. Impact of female sex hormones on the maturation and function of human dendritic cells. *Am. J. Reprod. Immunol.* **2009**, *62*, 165–173. [CrossRef]

32. Evans, M.J.; MacLaughlin, S.; Marvin, R.D.; Abdou, N.I. Estrogen decreases in vitro apoptosis of peripheral blood mononuclear cells from women with normal menstrual cycles and decreases TNF-alpha production in SLE but not in normal cultures. *Clin. Immunol. Immunopathol.* **1997**, *82*, 258–262. [CrossRef]

33. Karpuzoglu, E.; Zouali, M. The multi-faceted influences of estrogen on lymphocytes: Toward novel immuno-interventions strategies for autoimmunity management. *Clin. Rev. Allergy Immunol.* **2011**, *40*, 16–26. [CrossRef]

34. Pinzan, C.F.; Ruas, L.P.; Casabona-Fortunato, A.S.; Carvalho, F.C.; Roque-Barreira, M.C. Immunological basis for the gender differences in murine Paracoccidioides brasiliensis infection. *PLoS ONE* **2010**, *5*, e10757. [CrossRef]

35. Klein, P.W.; Easterbrook, J.D.; Lalime, E.N.; Klein, S.L. Estrogen and progesterone affect responses to malaria infection in female C57BL/6 mice. *Gend. Med.* **2008**, *5*, 423–433. [CrossRef]

36. Klein, S.L. Hormonal and immunological mechanisms mediating sex differences in parasite infection. *Parasite Immunol.* **2004**, *26*, 247–264. [CrossRef]

37. Gillgrass, A.E.; Fernandez, S.A.; Rosenthal, K.L.; Kaushic, C. Estradiol regulates susceptibility following primary exposure to genital herpes simplex virus type 2, while progesterone induces inflammation. *J. Virol.* **2005**, *79*, 3107–3116. [CrossRef]

38. Smith, S.M.; Baskin, G.B.; Marx, P.A. Estrogen protects against vaginal transmission of simian immunodeficiency virus. *J. Infect. Dis.* **2000**, *182*, 708–715. [CrossRef]

39. Pennock, J.W.; Stegall, R.; Bell, B.; Vargas, G.; Motamedi, M.; Milligan, G.; Bourne, N. Estradiol improves genital herpes vaccine efficacy in mice. *Vaccine* **2009**, *27*, 5830–5836. [CrossRef]

40. Sonoda, J.; Laganiere, J.; Mehl, I.R.; Barish, G.D.; Chong, L.W.; Li, X.; Scheffler, I.E.; Mock, D.C.; Bataille, A.R.; Robert, F.; et al. Nuclear receptor ERR alpha and coactivator PGC-1 beta are effectors of IFN-gamma-induced host defense. *Genes Dev.* **2007**, *21*, 1909–1920. [CrossRef]

41. Schneider, A.H.; Kanashiro, A.; Dutra, S.G.V.; Souza, R.D.N.; Veras, F.P.; Cunha, F.Q.; Ulloa, L.; Mecawi, A.S.; Reis, L.C.; Malvar, D.D.C. Estradiol replacement therapy regulates innate immune response in ovariectomized arthritic mice. *Int. Immunopharmacol.* **2019**, *72*, 504–510. [CrossRef]

42. Bouman, A.; Heineman, M.J.; Faas, M.M. Sex hormones and the immune response in humans. *Hum. Reprod. Update* **2005**, *11*, 411–423. [CrossRef]

43. Polanczyk, M.J.; Hopke, C.; Vandenbark, A.A.; Offner, H. Estrogen-mediated immunomodulation involves reduced activation of effector T cells, potentiation of Treg cells, and enhanced expression of the PD-1 costimulatory pathway. *J. Neurosci. Res.* **2006**, *84*, 370–378. [CrossRef]

44. Khan, D.; Dai, R.; Karpuzoglu, E.; Ahmed, S.A. Estrogen increases, whereas IL-27 and IFN-gamma decrease, splenocyte IL-17 production in WT mice. *Eur. J. Immunol.* **2010**, *40*, 2549–2556. [CrossRef]

45. Morrison, C.S.; Richardson, B.A.; Mmiro, F.; Chipato, T.; Celentano, D.D.; Luoto, J.; Mugerwa, R.; Padian, N.; Rugpao, S.; Brown, J.M.; et al. Hormonal contraception and the risk of HIV acquisition. *AIDS* **2007**, *21*, 85–95. [CrossRef]

46. Butts, C.L.; Shukair, S.A.; Duncan, K.M.; Bowers, E.; Horn, C.; Belyavskaya, E.; Tonelli, L.; Sternberg, E.M. Progesterone inhibits mature rat dendritic cells in a receptor-mediated fashion. *Int. Immunol.* **2007**, *19*, 287–296. [CrossRef]

47. Jones, L.A.; Kreem, S.; Shweash, M.; Paul, A.; Alexander, J.; Roberts, C.W. Differential modulation of TLR3- and TLR4-mediated dendritic cell maturation and function by progesterone. *J. Immunol.* **2010**, *185*, 4525–4534. [CrossRef]

48. Arruvito, L.; Giulianelli, S.; Flores, A.C.; Paladino, N.; Barboza, M.; Lanari, C.; Fainboim, L. NK cells expressing a progesterone receptor are susceptible to progesterone-induced apoptosis. *J. Immunol.* **2008**, *180*, 5746–5753. [CrossRef]

49. Piccinni, M.P.; Giudizi, M.G.; Biagiotti, R.; Beloni, L.; Giannarini, L.; Sampognaro, S.; Parronchi, P.; Manetti, R.; Annunziato, F.; Livi, C.; et al. Progesterone favors the development of human T helper cells producing Th2-type cytokines and promotes both IL-4 production and membrane CD30 expression in established Th1 cell clones. *J. Immunol.* **1995**, *155*, 128–133. [PubMed]

50. Gillgrass, A.E.; Ashkar, A.A.; Rosenthal, K.L.; Kaushic, C. Prolonged exposure to progesterone prevents induction of protective mucosal responses following intravaginal immunization with attenuated herpes simplex virus type 2. *J. Virol.* **2003**, *77*, 9845–9851. [CrossRef]

51. Kaushic, C.; Murdin, A.D.; Underdown, B.J.; Wira, C.R. Chlamydia trachomatis infection in the female reproductive tract of the rat: Influence of progesterone on infectivity and immune response. *Infect. Immun.* **1998**, *66*, 893–898. [CrossRef]

52. Marx, P.A.; Spira, A.I.; Gettie, A.; Dailey, P.J.; Veazey, R.S.; Lackner, A.A.; Mahoney, C.J.; Miller, C.J.; Claypool, L.E.; Ho, D.D.; et al. Progesterone implants enhance SIV vaginal transmission and early virus load. *Nat. Med.* **1996**, *2*, 1084–1089. [CrossRef] [PubMed]

53. Trunova, N.; Tsai, L.; Tung, S.; Schneider, E.; Harouse, J.; Gettie, A.; Simon, V.; Blanchard, J.; Cheng-Mayer, C. Progestin-based contraceptive suppresses cellular immune responses in SHIV-infected rhesus macaques. *Virology* **2006**, *352*, 169–177. [CrossRef] [PubMed]

54. Menzies, F.M.; Henriquez, F.L.; Alexander, J.; Roberts, C.W. Selective inhibition and augmentation of alternative macrophage activation by progesterone. *Immunology* **2011**, *134*, 281–291. [CrossRef] [PubMed]

55. Kamada, M.; Irahara, M.; Maegawa, M.; Yasui, T.; Takeji, T.; Yamada, M.; Tezuka, M.; Kasai, Y.; Aono, T. Effect of hormone replacement therapy on post-menopausal changes of lymphocytes and T cell subsets. *J. Endocrinol. Investig.* **2000**, *23*, 376–382. [CrossRef]

56. Ben-Hur, H.; Mor, G.; Insler, V.; Blickstein, I.; Amir-Zaltsman, Y.; Sharp, A.; Globerson, A.; Kohen, F. Menopause is associated with a significant increase in blood monocyte number and a relative decrease in the expression of estrogen receptors in human peripheral monocytes. *Am. J. Reprod. Immunol.* **1995**, *34*, 363–369. [CrossRef]

57. Engelmann, F.; Rivera, A.; Park, B.; Messerle-Forbes, M.; Jensen, J.T.; Messaoudi, I. Impact of Estrogen Therapy on Lymphocyte Homeostasis and the Response to Seasonal Influenza Vaccine in Post-Menopausal Women. *PLoS ONE* **2016**, *11*, e0149045. [CrossRef]

58. Yang, J.H.; Chen, C.D.; Wu, M.Y.; Chao, K.H.; Yang, Y.S.; Ho, H.N. Hormone replacement therapy reverses the decrease in natural killer cytotoxicity but does not reverse the decreases in the T-cell subpopulation or interferon-gamma production in postmenopausal women. *Fertil. Steril.* **2000**, *74*, 261–267. [CrossRef]

59. Fahlman, M.M.; Boardley, D.; Flynn, M.G.; Bouillon, L.E.; Lambert, C.P.; Braun, W.A. Effects of hormone replacement therapy on selected indices of immune function in postmenopausal women. *Gynecol. Obstet. Investig.* **2000**, *50*, 189–193. [CrossRef]

60. Porter, V.R.; Greendale, G.A.; Schocken, M.; Zhu, X.; Effros, R.B. Immune effects of hormone replacement therapy in post-menopausal women. *Exp. Gerontol.* **2001**, *36*, 311–326. [CrossRef]

61. Kamada, M.; Irahara, M.; Maegawa, M.; Yasui, T.; Yamano, S.; Yamada, M.; Tezuka, M.; Kasai, Y.; Deguchi, K.; Ohmoto, Y.; et al. B cell subsets in postmenopausal women and the effect of hormone replacement therapy. *Maturitas* **2001**, *37*, 173–179. [CrossRef]

62. Kumru, S.; Godekmerdan, A.; Yilmaz, B. Immune effects of surgical menopause and estrogen replacement therapy in peri-menopausal women. *J. Reprod. Immunol.* **2004**, *63*, 31–38. [CrossRef]

63. Holl, M.; Donat, H.; Weise, W. Peripheral blood lymphocyte subpopulations of postmenopausal women with hormone replacement therapy. *Zentralbl. Gynakol.* **2001**, *123*, 543–545.

64. Harlow, S.D.; Gass, M.; Hall, J.E.; Lobo, R.; Maki, P.; Rebar, R.W.; Sherman, S.; Sluss, P.M.; de Villiers, T.J.; Group, S.C. Executive summary of the Stages of Reproductive Aging Workshop + 10: Addressing the unfinished agenda of staging reproductive aging. *Menopause* **2012**, *19*, 387–395. [CrossRef]

65. Kim, O.Y.; Chae, J.S.; Paik, J.K.; Seo, H.S.; Jang, Y.; Cavaillon, J.M.; Lee, J.H. Effects of aging and menopause on serum interleukin-6 levels and peripheral blood mononuclear cell cytokine production in healthy nonobese women. *Age* **2012**, *34*, 415–425. [CrossRef]

66. Giuliani, N.; Sansoni, P.; Girasole, G.; Vescovini, R.; Passeri, G.; Passeri, M.; Pedrazzoni, M. Serum interleukin-6, soluble interleukin-6 receptor and soluble gp130 exhibit different patterns of age- and menopause-related changes. *Exp. Gerontol.* **2001**, *36*, 547–557. [CrossRef]

67. Brooks-Asplund, E.M.; Tupper, C.E.; Daun, J.M.; Kenney, W.L.; Cannon, J.G. Hormonal modulation of interleukin-6, tumor necrosis factor and associated receptor secretion in postmenopausal women. *Cytokine* **2002**, *19*, 193–200. [CrossRef]

68. Saucedo, R.; Rico, G.; Basurto, L.; Ochoa, R.; Zarate, A. Transdermal estradiol in menopausal women depresses interleukin-6 without affecting other markers of immune response. *Gynecol. Obstet. Investig.* **2002**, *53*, 114–117. [CrossRef]

69. Vural, P.; Canbaz, M.; Akgul, C. Effects of menopause and postmenopausal tibolone treatment on plasma TNFalpha, IL-4, IL-10, IL-12 cytokine pattern and some bone turnover markers. *Pharmacol. Res.* **2006**, *53*, 367–371. [CrossRef]

70. Yasui, T.; Maegawa, M.; Tomita, J.; Miyatani, Y.; Yamada, M.; Uemura, H.; Matsuzaki, T.; Kuwahara, A.; Kamada, M.; Tsuchiya, N.; et al. Changes in serum cytokine concentrations during the menopausal transition. *Maturitas* **2007**, *56*, 396–403. [CrossRef] [PubMed]

71. Abu-Taha, M.; Rius, C.; Hermenegildo, C.; Noguera, I.; Cerda-Nicolas, J.M.; Issekutz, A.C.; Jose, P.J.; Cortijo, J.; Morcillo, E.J.; Sanz, M.J. Menopause and ovariectomy cause a low grade of systemic inflammation that may be prevented by chronic treatment with low doses of estrogen or losartan. *J. Immunol.* **2009**, *183*, 1393–1402. [CrossRef] [PubMed]

72. Da Silva, L.H.; Panazzolo, D.G.; Marques, M.F.; Souza, M.G.; Paredes, B.D.; Nogueira Neto, J.F.; Leao, L.M.; Morandi, V.; Bouskela, E.; Kraemer-Aguiar, L.G. Low-dose estradiol and endothelial and inflammatory biomarkers in menopausal overweight/obese women. *Climacteric.* **2016**, *19*, 337–343. [CrossRef] [PubMed]

73. Stopinska-Gluszak, U.; Waligora, J.; Grzela, T.; Gluszak, M.; Jozwiak, J.; Radomski, D.; Roszkowski, P.I.; Malejczyk, J. Effect of estrogen/progesterone hormone replacement therapy on natural killer cell cytotoxicity and immunoregulatory cytokine release by peripheral blood mononuclear cells of postmenopausal women. *J. Reprod. Immunol.* **2006**, *69*, 65–75. [CrossRef]

74. Calippe, B.; Douin-Echinard, V.; Delpy, L.; Laffargue, M.; Lelu, K.; Krust, A.; Pipy, B.; Bayard, F.; Arnal, J.F.; Guery, J.C.; et al. 17Beta-estradiol promotes TLR4-triggered proinflammatory mediator production through direct estrogen receptor alpha signaling in macrophages in vivo. *J. Immunol.* **2010**, *185*, 1169–1176. [CrossRef]

75. Vegeto, E.; Ghisletti, S.; Meda, C.; Etteri, S.; Belcredito, S.; Maggi, A. Regulation of the lipopolysaccharide signal transduction pathway by 17beta-estradiol in macrophage cells. *J. Steroid Biochem. Mol. Biol.* **2004**, *91*, 59–66. [CrossRef]

76. Ghisletti, S.; Meda, C.; Maggi, A.; Vegeto, E. 17beta-estradiol inhibits inflammatory gene expression by controlling NF-kappaB intracellular localization. *Mol. Cell. Biol.* **2005**, *25*, 2957–2968. [CrossRef]

77. Lamas, A.Z.; Caliman, I.F.; Dalpiaz, P.L.; de Melo, A.F., Jr.; Abreu, G.R.; Lemos, E.M.; Gouvea, S.A.; Bissoli, N.S. Comparative effects of estrogen, raloxifene and tamoxifen on endothelial dysfunction, inflammatory markers and oxidative stress in ovariectomized rats. *Life Sci.* **2015**, *124*, 101–109. [CrossRef]

78. De Medeiros, A.R.; Lamas, A.Z.; Caliman, I.F.; Dalpiaz, P.L.; Firmes, L.B.; de Abreu, G.R.; Moyses, M.R.; Lemos, E.M.; dos Reis, A.M.; Bissoli, N.S. Tibolone has anti-inflammatory effects in estrogen-deficient female rats on the natriuretic peptide system and TNF-alpha. *Regul. Pept.* **2012**, *179*, 55–60. [CrossRef]

79. Ververidis, F.; Trantas, E.; Douglas, C.; Vollmer, G.; Kretzschmar, G.; Panopoulos, N. Biotechnology of flavonoids and other phenylpropanoid-derived natural products. Part I: Chemical diversity, impacts on plant biology and human health. *Biotechnol. J.* **2007**, *2*, 1214–1234. [CrossRef]

80. Ryan-Borchers, T.A.; Park, J.S.; Chew, B.P.; McGuire, M.K.; Fournier, L.R.; Beerman, K.A. Soy isoflavones modulate immune function in healthy postmenopausal women. *Am. J. Clin. Nutr.* **2006**, *83*, 1118–1125. [CrossRef]

81. Rachon, D.; Rimoldi, G.; Wuttke, W. In vitro effects of genistein and resveratrol on the production of interferon-gamma (IFNgamma) and interleukin-10 (IL-10) by stimulated murine splenocytes. *Phytomedicine* **2006**, *13*, 419–424. [CrossRef]

82. Zhang, Y.; Song, T.T.; Cunnick, J.E.; Murphy, P.A.; Hendrich, S. Daidzein and genistein glucuronides in vitro are weakly estrogenic and activate human natural killer cells at nutritionally relevant concentrations. *J. Nutr.* **1999**, *129*, 399–405. [CrossRef]

83. Guo, T.L.; McCay, J.A.; Zhang, L.X.; Brown, R.D.; You, L.; Karrow, N.A.; Germolec, D.R.; White, K.L., Jr. Genistein modulates immune responses and increases host resistance to B16F10 tumor in adult female B6C3F1 mice. *J. Nutr.* **2001**, *131*, 3251–3258. [CrossRef]

84. Baeza, I.; De Castro, N.M.; Gimenez-Llort, L.; De la Fuente, M. Ovariectomy, a model of menopause in rodents, causes a premature aging of the nervous and immune systems. *J. Neuroimmunol.* **2010**, *219*, 90–99. [CrossRef]

85. Askoxylaki, M.; Siristatidis, C.; Chrelias, C.; Vogiatzi, P.; Creatsa, M.; Salamalekis, G.; Vrantza, T.; Vrachnis, N.; Kassanos, D. Reactive oxygen species in the follicular fluid of subfertile women undergoing In Vitro Fertilization: A short narrative review. *J. Endocrinol. Investig.* **2013**, *36*, 1117–1120.

86. Chi, X.X.; Zhang, T. The effects of soy isoflavone on bone density in north region of climacteric Chinese women. *J. Clin. Biochem. Nutr.* **2013**, *53*, 102–107. [CrossRef]

87. Nadadur, M.; Stanczyk, F.Z.; Tseng, C.C.; Kim, L.; Wu, A.H. The Effect of Reduced Dietary Fat and Soy Supplementation on Circulating Adipocytokines in Postmenopausal Women: A Randomized Controlled 2-Month Trial. *Nutr. Cancer* **2016**, *68*, 554–559. [CrossRef] [PubMed]

88. Llaneza, P.; Gonzalez, C.; Fernandez-Inarrea, J.; Alonso, A.; Diaz, F.; Arnott, I.; Ferrer-Barriendos, J. Soy isoflavones, diet and physical exercise modify serum cytokines in healthy obese postmenopausal women. *Phytomedicine.* **2011**, *18*, 245–250. [CrossRef] [PubMed]

89. Owor, R.O.; Bedane, K.G.; Openda, Y.I.; Zuhlke, S.; Derese, S.; Ong'amo, G.; Ndakala, A.; Spiteller, M. Synergistic anti-inflammatory activities of a new flavone and other flavonoids from *Tephrosia hildebrandtii* vatke. *Nat. Prod. Res.* **2020**, 1–8. [CrossRef] [PubMed]

90. Owor, R.O.; Derese, S.; Bedane, K.G.; Zuhlke, S.; Ndakala, A.; Spiteller, M. Isoflavones from the seedpods of Tephrosia vogelii and pyrazoisopongaflavone with anti-inflammatory effects. *Fitoterapia.* **2020**, *146*, 104695. [CrossRef]

91. Beavers, K.M.; Serra, M.C.; Beavers, D.P.; Cooke, M.B.; Willoughby, D.S. Soymilk supplementation does not alter plasma markers of inflammation and oxidative stress in postmenopausal women. *Nutr. Res.* **2009**, *29*, 616–622. [CrossRef]

92. Huang, Y.; Cao, S.; Nagamani, M.; Anderson, K.E.; Grady, J.J.; Lu, L.J. Decreased circulating levels of tumor necrosis factor-alpha in postmenopausal women during consumption of soy-containing isoflavones. *J. Clin. Endocrinol. Metab.* **2005**, *90*, 3956–3962. [CrossRef]

93. Giolo, J.S.; Costa, J.G.; da Cunha-Junior, J.P.; Pajuaba, A.; Taketomi, E.A.; de Souza, A.V.; Caixeta, D.C.; Peixoto, L.G.; de Oliveira, E.P.; Everman, S.; et al. The Effects of Isoflavone Supplementation Plus Combined Exercise on Lipid Levels, and Inflammatory and Oxidative Stress Markers in Postmenopausal Women. *Nutrients* **2018**, *10*, 424. [CrossRef]

Article

Deficiency in the Essential Amino Acids L-Isoleucine, L-Leucine and L-Histidine and Clinical Measures as Predictors of Moderate Depression in Elderly Women: A Discriminant Analysis Study

Silvia Solís-Ortiz [1,*], Virginia Arriaga-Avila [2], Aurora Trejo-Bahena [1] and Rosalinda Guevara-Guzmán [2]

[1] Departamento de Ciencias Médicas, División de Ciencias de la Salud, Campus León, Universidad de Guanajuato, León, Guanajuato 37320, Mexico; aurora.st@gmail.com

[2] Facultad de Medicina, Universidad Nacional Autónoma de México, Ciudad de México 04510, Mexico; varriaga@servidor.unam.mx (V.A.-A.); rguevara@servidor.unam.mx (R.G.-G.)

* Correspondence: ms.solis@ugto.mx; Tel./Fax: +52-477-7145859

Abstract: Increases in depression are common in some elderly women. Elderly women often show moderate depressive symptoms, while others display minimal depressive symptoms. These discrepancies have produced contradictory and inconclusive outcomes, which have not been explained entirely by deficits in neurotransmitter precursors. Deficiency in some amino acids have been implicated in major depression, but its role in non-clinical elderly women is not well known. An analysis of essential amino acids, depression and the use of discriminant analysis can help to clarify the variation in depressive symptoms exhibited by some elderly women. The aim was to investigate the relationship of essential amino acids with affective, cognitive and comorbidity measures in elderly women without major depression nor severe mood disorders or psychosis, specifically thirty-six with moderate depressive symptoms and seventy-one with minimal depressive symptoms. The plasma concentrations of nineteen amino acids, Beck Depression Inventory (BDI) scores, Geriatric Depression Scale (GDS) scores, global cognitive scores and comorbidities were submitted to stepwise discriminant analysis to identify predictor variables. Seven predictors arose as important for belong to the group based on amino acid concentrations, with the moderate depressive symptoms group characterized by higher BDI, GDS and cognitive scores; fewer comorbidities; and lower levels of L-histidine, L-isoleucine and L-leucine. These findings suggest that elderly women classified as having moderate depressive symptoms displayed a deficiency in essential amino acids involved in metabolism, protein synthesis, inflammation and neurotransmission.

Keywords: depression; cognition; symptoms; elderly; amino acids; comorbidities

Citation: Solís-Ortiz, S.; Arriaga-Avila, V.; Trejo-Bahena, A.; Guevara-Guzmán, R. Deficiency in the Essential Amino Acids L-Isoleucine, L-Leucine and L-Histidine and Clinical Measures as Predictors of Moderate Depression in Elderly Women: A Discriminant Analysis Study. *Nutrients* **2021**, *13*, 3875. https://doi.org/10.3390/nu13113875

Academic Editor: Masakazu Terauchi

Received: 14 July 2021
Accepted: 25 July 2021
Published: 29 October 2021

Publisher's Note: MDPI stays neutral with regard to jurisdictional claims in published maps and institutional affiliations.

Copyright: © 2021 by the authors. Licensee MDPI, Basel, Switzerland. This article is an open access article distributed under the terms and conditions of the Creative Commons Attribution (CC BY) license (https://creativecommons.org/licenses/by/4.0/).

1. Introduction

According to the World Health Organization [1], depressive symptoms include feelings of sadness, hopelessness, pessimism, and low self-esteem; a decrease in or loss of ability to feel pleasure; reduced energy and vitality; slowness of thought; loss of appetite and disturbed sleep or insomnia. These depressive symptoms are prevalent among elderly individuals [2], with a higher incidence in elderly women [3]. In some elderly women, the symptoms extend from mild to severe and need medication [4], whereas in other elderly women, symptoms are minor or absent, maybe representing the common depressive symptoms of typical elderly women [5]. Affective studies have demonstrated that for some elderly women, severe symptoms of depression may be difficult to recognize because they may exhibit different symptoms than in younger people [1]. For some elderly women with depression, sadness is not their main symptom, they may have other, less obvious symptoms of depression, probably related to other comorbidities and cognitive impairment [6]. Depression in elderly individuals is multifactorial, which makes its assessment difficult [7];

however, deficiencies in some amino acids involved in brain metabolism can be a relevant factor for alterations in mood [8]. Some studies have suggested an association between plasma amino acids and psychosis in mixed populations [9]. Individuals with the first episode of psychosis had reduced γ-aminobutyric acid (GABA) plasma levels [10] and decreased levels of proline, alpha-aminoadipic acid, kynurenine, valine, tyrosine, citrulline, tryptophan, and histidine compared to controls [11]. Another study found low levels of tryptophan associated with major depression [12]. Some studies have also reported low levels of tryptophan, methionine, phenylalanine, and tyrosine [13,14], as well as GABA, dopamine, tyramine and kynurenine in patients with major depression compared to controls [15]. Branched-chain amino acids have been implicated in major depression, although they have not been extensively analyzed. Low levels of leucine were reported in patients diagnosed with bipolar disorder [16], whereas leucine, valine and isoleucine levels were decreased in middle-aged patients of both genders with major depression [17]. However, most results related to biomarkers are inconsistent due to the heterogeneity of depression and the diverse methodology used, such as the inclusion of individuals of different ages, sexes and diagnoses [18].

Depressive symptoms are not sufficiently characterized in elderly women over 65 years of age because there are elderly women who show moderate depressive symptoms, while others display minimal depressive symptoms. According to the American Psychiatric Association, minor depressive symptoms are characterized by symptoms that often pass unnoticed because the presence of depressed mood is not detected, nor the inability to enjoy things. In contrast, moderate depressive symptoms are characterized by the manifestation of symptoms that produce some difficulty in continuing with activities of daily living, whose intensity may cause discomfort but is manageable. These discrepancies have produced ambiguous outcomes, which have not been resolved completely by factors related to depression. Most studies have used correlation analysis to identify relationships between depressive scores and declining amino acid levels in individuals of different ages who were diagnosed with major depression. In some investigations, this approximation has provided appropriate understanding, but other investigations have informed conflicting outcomes. The interpretation of correlation analysis is suitable but limited when one demands to detect the features that differentiate two groups and to establish a function capable of discern between the members of the two groups with the large feasible accuracy. Discriminant analysis allows the prediction of group membership from a set of predictors (independent variables) separating these variables from others that are orthogonally independent [19]; hence, discriminant analysis is an appropriate statistical method to detect the variables that allow differentiation between groups and to establish how many of these variables are required to reach the optimal feasible categorization [20]. An examination of the discriminant function coefficients of amino acid profiles and affective, global cognition and comorbidity measures between moderately and minimally depressed women, undiagnosed with major depressive disorder, by discriminant analysis may help us in enhancing our comprehension of the impact of amino acid deficiencies on depressive symptoms.

The aim of the present research was to examine the amino acid profiles and the mood, global cognitive and comorbidity measures in elderly women with moderate and minimal levels of depressive symptoms, to identify, through discriminant analysis, predictor variables that best classify subjects. This classification could help to elucidate the variation in depressive symptoms shown by some elderly females related with optimal or deficiencies in amino acid concentrations. We expected that the moderate depressive symptoms group would exhibit profile variables that could reflect alterations in mood based on depressive symptom scores, global cognitive scores, the number of comorbidities and plasma amino acid profiles.

2. Materials and Methods

2.1. Participants

Ninety-seven elderly women between 65 and 79 years of age participated in this cross-sectional study. The elderly women were recruited from a senior center. This center provides health care to elderly women through recreational and occupational activities. The study only included a sample of elderly women without signs of pre-existing severe mental disorders, according to the Diagnostic and Statistical Manual of Mental Disorders, 5th Edition (DSM-5) [21]. Women with severe major depression, bipolar disorder, dysthymia, dysphoria or psychosis such as schizophrenia were excluded from the study. Therefore, this is a non-clinical sample that describes the depressive symptoms of a group of elderly women attending a senior center. To estimate the statistical power of the sample size of 97 women, the mean and standard deviations were obtained. The BDI [22] depressive scores for each group were used to separate the patients into two groups. A power of 0.80 was regarded to identify a 10% contrast in BDI scores, with a significance level of $\alpha = 0.05$. The power of this sample size was appropriate to continue with the examination of the variables of the current research, with a large effect size (Cohen's d = 0.83). Since females show more susceptibility to depression [23], only females were selected for this investigation. The elderly women were divided into a minimal depressive symptoms group ($n = 61$) and a moderate depressive symptoms group ($n = 36$) according to their scores of depressive symptoms on the BDI questionnaire. So, the two groups were contrasted. The stratification criterion was based on these punctuations: 61 elderly women who scored between 0 and 9 were classified in the minimal depressive symptoms group, whereas 36 elderly women who scored between 10 and 29 were classified in the moderate depressive symptoms group. Thus, a high score was indicative of moderate depressive symptoms [5]. The elderly women were examined in a unique session (between 0900 h and 1000 h) by one trained investigator, who asked them about their additional illnesses, medications and hormone replacement therapy. The elderly women included in the study were not treated with antidepressant. This research was approved by the Research and Ethics Committee of the Faculty of Medicine of the National Autonomous University of Mexico (No. 007-2018). All of the elderly females conceded written informed consent before to participate in the investigation.

2.2. Depression Questionnaires

The BDI questionnaire [22], in a standardized version for the Mexican population [24], was used to evaluate self-reports of current depressive symptoms in 97 elderly women. The BDI questionnaire consists of 21 items that measure current depressive symptoms. Each item contains a group of four statements, from which the subject chooses one according to how she felt in the last week. These statements reflect the severity of the discomfort produced by depressive symptoms and are marked from 0 (minimal) to 4 (severe). The total score of scale is obtained by adding the scores for the 21 items, with 0 as the lowest score and 64 as the maximum score. Individual questions of the BDI assess mood, pessimism, sense of failure, self-dissatisfaction, guilt, punishment, self-dislike, self-accusation, suicidal ideas, crying, irritability, social withdrawal, body image, work difficulties, insomnia, fatigue, appetite, weight loss, bodily preoccupation, and loss of libido. Items 1 to 13 assess symptoms that are psychological in nature, while items 14 to 21 assess more physical symptoms. In the present study, we used the cut-off points to categorize the severity of depression according to the version of the BDI questionnaire adapted for the Mexican population [24]. A score of 0 to 9 reflected the absence or minimal presence of depressive symptoms; 10 to 16 indicated a medium depression; 17 to 29 reflected moderate depression and scores of 30 to 63 indicated severe depression. Since only one elderly woman scored in medium depression, she was aggregated with the elderly women who scored in the 17 to 29 range, so scores from 10 to 29 were considered moderate depressive symptoms. Furthermore, one of the participants who scored in the range of 30 to 35 was excluded from the study. Depressive scores were compared between groups with minimal and moderate symptoms.

The GDS questionnaire [25], in a standardized version for the Spanish-speaking population [26], was also used to identify symptoms of depression in elderly women. The GDS questionnaire evaluates sadness, lack of energy, positive mood, social withdrawal, agitation, feelings of worthlessness, feelings of despair and perception of cognitive changes. Each item on the GDS is scored with one point for a depressive response. Items are summed to determine the total score with a maximum of 30 points. Scores ranging from 0 to 9 indicate normal mood, scores of 10 to 19 indicate mild depressive symptoms, and scores of 20 to 30 indicate severe depressive symptoms.

2.3. Dementia and Global Cognition

Initial dementia and general cognition were assessed employing the Mini-Mental State Examination (MMSE) [27] in a standardized version for the Spanish-speaking population [28] to establish cognitive status in different areas that can be related to different cognitive symptoms. These cognitive symptoms include temporal and spatial orientation, immediate memory and retention, concentration and working memory, language and graphic constructive praxis. Possible scores on this examination range from 0 to 30, and subjects with dementia usually score under 24.

2.4. Quantification of Amino Acids

Before the depressive symptom evaluation period, a 10 mL blood sample was obtained from an antecubital vein from the women between 0800 h and 0900 h after a nocturne fast. Blood samples were centrifuged at 25 °C to get plasma. The plasma was accumulated, distributed into 2 mL tubes, and stored at −80 °C until examination. The quantification of the amino acids in the plasma was performed using an Agilent 1100 HPLC system (Agilent Technologies, Santa Clara, CA, USA) based on the procedure reported by Henderson et al. [29]. The following 19 amino acids were quantified with this method: L-aspartate, L-glutamate, L-asparagine, L-serine, L-glutamine, L-histidine, L-glycine, L-threonine, L-citrulline, L-arginine, L-alanine + taurine, GABA, L-tyrosine, L-valine, L-methionine, L-tryptophan, L-phenylalanine, L-isoleucine and L-leucine.

2.5. Discriminant Analysis

The current investigation executed stepwise discriminant analysis [19] to create a model to prognosticate which groups the subjects belong to. Also, to determine predictor variables of moderate and minimal depressive symptoms in elderly women based on their BDI depressive scores, GDS depressive scores, MMSE scores, comorbidities and amino acid profiles. We employed discriminant analysis because we were focused in detecting prognostic variables of group membership [30]; therefore, the discriminant analysis was executed after the females were divided into moderate and minimal depressive symptom groups.

2.6. Statistical Procedure

Statistical procedure and the power of the sample were determined with the Statistical Package for the Social Sciences (SPSS) version 23, IBM, New York, NY, USA. We used a two-tailed α value of 0.05 and a β value of 0.80 to calculate a standardized effect size of 10% based on the BDI scores between groups. Prior to statistical procedures were employed, the data were inspected for distribution normality applying the Kolmogorov-Smirnov test. Partial Wilks' lambda was used to compare the means of the characteristics of the elderly women and their amino acid profiles between the two groups. A stepwise multiple discriminant analysis [19] was executed to detect latent variables as predictors of moderate and minimal depressive symptoms in elderly women based on their BDI depressive scores, GDS depressive scores, MMSE scores, comorbidities, and plasma concentrations of amino acids. Contrasts were considered as significant when p was <0.05.

3. Results

3.1. Characteristics of the Participants and Depressive Symptoms

Sample characteristics included age, years of schooling, age at menarche, age at menopause, weight, height, body mass index, systolic arterial pressure, diastolic arterial pressure, comorbidities, BDI depression scores, GDS depression scores and MMSE scores, which are presented in Table 1. The results show the test of equality of group means with the coefficients of the partial Wilks' lambda derived from the discriminant analysis. The BDI and GDS depression scores were significantly elevated in the moderate-depressive group, which indicated that this group scored between 10 and 29 on the BDI and between 9 and 14 on the GDS, respectively. None of the elderly women had severe depressive symptoms. There were no noteworthy contrasts among the minimal and moderate depressive symptoms groups in the rest of their features or in their dementia indexes or comorbidities. The most prevalent comorbidities reported by the elderly women included hypertension (63%), osteoporosis (69%), arthritis (52%), gynecological disease (78%), surgery (94%) and kidney disease (67%). Other diseases (62%) included glaucoma, migraine, hepatitis, hypothyroidism, diabetes, allergies and sinusitis.

Table 1. Features of the elderly women with minimal and moderate depressive symptoms.

Features	Minimal Mean $\pm$ SD	Moderate Mean $\pm$ SD	Partial Wilks' Lambda	*F* Value	*p* Value
Age (years)	66 $\pm$ 6.0	67 $\pm$ 5.7	0.993	0.67	0.41
Years of education	8 $\pm$ 5.6	6 $\pm$ 4.0	0.976	2.35	0.12
Age at Menarche	12 $\pm$ 1.0	13 $\pm$ 1.0	0.992	0.958	0.35
Age at Menopause	48 $\pm$ 3.9	47 $\pm$ 4.0	0.998	-1.022	0.25
Weight (kg)	72 $\pm$ 13.0	68 $\pm$ 7.0	0.981	-0.426	0.10
Size (m)	1.5 $\pm$ 0.10	1.5 $\pm$ 0.10	0.978	-0.805	0.92
BMI (kg/m^2)	30 $\pm$ 5.0	28 $\pm$ 3.0	0.973	0.112	0.91
TAS (mmHg)	113 $\pm$ 11.0	127 $\pm$ 9.0	0.992	0.335	0.40
TAD (mmHg)	70 $\pm$ 6.0	75 $\pm$ 5.0	0.966	0.633	0.08
Comorbidities	2 $\pm$ 0.9	2 $\pm$ 1.0	0.998	0.22	0.64
GDS depression scores	3 $\pm$ 3.6	9 $\pm$ 5.7	0.715	38.20	0.0001
BDI depression scores	4.5 $\pm$ 2.6	18 $\pm$ 6.6	0.331	194.00	0.0001
MMSE scores	24.5 $\pm$ 4.3	26 $\pm$ 3.1	0.987	1.24	0.27

MMSE = Mini Mental State Examination; BMI = Body Mass Index; TAS = Systolic arterial tension; TAD = Diastolic arterial tension.

3.2. Amino Acid Profiles

The plasma amino acid profiles of the minimal and moderate depressive symptom groups are presented in Table 2. The results show the test of equality of group means with the coefficients of the partial Wilks' lambda derived from the discriminant analysis. Plasma concentrations of L-methionine were elevated in the moderate depressive symptom groups, while L-isoleucine levels were elevated in the minimal depressive symptom groups. L-Aspartate, L-glutamate, L-asparagine, L-serine, L-glutamine, L-histidine, L-glycine, L-threonine, L-citrulline, L-arginine, L-alanine + taurine, GABA, L-tyrosine, L-valine, L-methionine, L-tryptophan, L-phenylalanine and L-leucine did not display notable contrasts between the moderate and minimal depressive symptom groups. Discriminant analysis revealed additional elements, which are exposed below.

Table 2. Amino-acid profiles (µM) of the elderly women with minimal and moderate depressive symptoms.

Amino-Acids	Minimal Mean ± SD	Moderate Mean ± SD	Partial Wilks' Lambda	*F* Value	*p* Value
L-Aspartate	3.2 ± 4.2	4.2 ± 3.9	0.986	1.367	0.24
L-Glutamate	166.6 ± 153.2	173.6 ± 90.5	0.999	0.066	0.79
L-Asparagine	27.2 ± 7.0	26.6 ± 7.0	0.998	0.157	0.69
L-Serine	105.3 ± 24.8	110.2 ± 18.4	0.988	1.164	0.28
L-Glutamine	383.2 ± 139.2	361.8 ± 102.3	0.993	0.656	0.40
L-Histidine	62.0 ± 10.8	59.8 ± 8.3	0.988	1.158	0.28
L-Glycine	272.6 ± 74.2	251.9 ± 72.0	0.981	1.825	0.18
L-Theonine	104.9 ± 24.4	107.5 ± 28.8	0.998	0.237	0.62
L-Cirtrulline	23.8 ± 6.5	25.2 ± 7.1	0.990	1.007	0.31
L-Arginine	48.0 ± 21.3	50.7 ± 20.0	0.996	0.386	0.53
L-Alanine + Taurine	195.4 ± 45.3	192.8 ± 39.8	0.999	0.086	0.77
GABA	2.2 ± 2.0	2.6 ± 2.8	0.991	0.916	0.34
L-Tyrosine	58.6 ± 12.7	63.1 ± 16.9	0.979	2.009	0.16
L-Valine	182.8 ± 27.8	192.8 ± 32.3	0.973	2.635	0.10
L-Methionine	25.3 ± 8.0	28.7 ± 9.7	0.966	3.429	0.05
L-Tryptophan	32.7 ± 7.1	34.5 ± 9.7	0.984	1.552	0.21
L-Phenylalanine	51.5 ± 8.5	51.5 ± 8.1	1.000	0.002	0.96
L-Isoleucine	55.9 ± 14.5	49.3 ± 11.7	0.939	6.232	0.01
L-Leucine	123.0 ± 22.0	119.5 ± 20.2	0.993	0.651	0.42

GABA = γ-aminobutyric acid

3.3. Discriminant Analysis

The discriminant analysis results are displayed in Table 3. Discriminant analysis detected seven latent variables as predictors of moderate and minimal depressive symptoms: BDI scores, GDS scores, MMSE scores, comorbidities, L-histidine, L-isoleucine, L-leucine and one discriminant function, which described 100% of the variance, canonical = 0.864. The canonical correlation value was squared to estimate the effect size of the discriminant function, $R^2 = 0.74$, which showed a large effect size. This discriminant function significantly differentiated the depressive groups, Wilks's lambda = 0.253; $\chi^2 = 127.20$, df = 7, $p < 0.001$. The small lambda indicates that the group means diverge. The eigenvalue was great at 2.956, showing that the function differentiates well among the groups, and the canonical correlation was near to 1 (0.86). Standardized canonical discriminant function coefficients showed that BDI scores, GDS scores, MMSE scores, comorbidities and L-isoleucine had a positive relationship, whereas L-histidine and L-leucine had a negative relationship with this function. High punctuations demonstrate that a dependent variable is crucial for a variate, and variables with positive and negative coefficients contribute to the variate in opposite ways [30]. Structure matrix coefficients showed that BDI scores (0.82) and GDS scores (0.36) had high canonical correlations and contributed the most to group separation, whereas L-isoleucine (0.14), MMSE scores (0.06), L-leucine (0.04) and comorbidities (0.02) contributed less and L-histidine (−0.06) contributed in the opposite manner. High canonical variate correlations indicate that these variables contribute the most to group separations.

Table 3. Outcomes of standardized canonical discriminant function coefficients and predictor variables ordered by absolute size of correlation within function.

Predictor Variables	Standardized Coefficients	Structure Matrix	*F* Ratio	*p* Value
BDI depression scores	0.961	0.827	194.00	0.0001
MMSE scores	0.386	0.066	102.35	0.0001
Comorbidities	0.396	0.028	74.60	0.0001
L-Histidine	−0.196	−0.064	58.87	0.0001
GDS depression scores	0.333	0.367	49.34	0.0001
L-Isoleucine	0.508	0.148	42.39	0.0001
L-Leucine	−0.396	0.048	38.00	0.0001
Canonical correlation	0.864			
Effect size R^2	0.74			
Eigenvalue	2.956			
Wilks' lambda	0.253			
x^2	127.20			0.0001

GDS = Geriatric Depression Scale; BDI = Beck Depression Inventory; MMSE = Mini Mental State Examination.

The estimate of the functions at group centroids distinguished the centroids of the moderate-depressive symptoms group, which were situated in the positive zone (2.185), whereas the centroid of the minimal-depressive symptoms group was situated in the negative zone (−1.325). These outcomes showed that the groups with values opposite in sign were discriminated by the functions. Of the 61 elderly women who were in the minimal-depressive symptoms group, 61 (100%) were correctly classified as members of that group, and the 36 who were in the moderate-depressive symptoms group, 36 were also correctly classified supported on the selected variables.

The outcomes of classification function coefficients in moderate- and minimal-depressive symptom groups supported on Fisher´s linear discriminant functions are displayed in Table 4. These coefficients mention us between which of the groups the specific functions discriminate and permit us to establish to which group each subject most probable belongs. These coefficients revealed that the moderate-depressive symptoms group displayed high coefficients in the predictor variables, which suggests that this group had higher BDI, GDS, and MMSE scores; fewer comorbidities; and reduced levels of L-histidine, L-isoleucine, and L-leucine contrasted to the minimal-depressive symptoms group.

Table 4. Outcomes of classification function coefficients in moderate and minimal depressive symptom groups.

Predictors	Moderate Symptoms	Minimal Symptoms
BDI depression scores	0.579	−0.157
GDS depression scores	1.201	0.943
MMSE scores	2.243	1.903
Comorbidities	0.276	1.658
L-histidine	0.365	0.434
L-isoleucine	0.019	−0.120
L-leucine	0.220	0.286

Fisher´s linear discrimination functions

Our results showed, through the structure matrix, that the coefficients with high and low canonical correlations were used to establish group membership. They seem to indicate some associations. However, a causality cannot be established with this analysis, so future studies should consider the inclusion of other analysis, methods and variables.

4. Discussion

The current study identified seven factors as significant predictors of group membership based on amino acid concentrations, depression scores and global cognitive scores, with the moderate depressive symptoms group characterized by higher BDI, GDS scores, and cognitive scores; fewer comorbidities; and lower levels of L-histidine, L-isoleucine and L-leucine. These results indicate that elderly women classified in the moderate depressive symptoms group exhibited deficiencies in essential amino acids in addition to self-report of depressive symptoms. Our outcomes also revealed that patients with minimal depressive symptoms had optimal concentrations of essential amino acids. This could explain, at least in part, the moderate depressive symptoms exhibited by some elderly females. Moreover, discriminant analysis successfully classified group membership; hence, the differences between the groups could be established. The present study did not perform additional analyzes to find definitive causal relationships between essential amino acids and depression. However, the discriminant analysis provides a multivariate analysis that include χ^2, canonical correlations, and through the structure matrix, the coefficients with high and low canonical correlations are obtained and used to establish group membership associations [19], based on the essential amino acids and depression.

The present study showed that BDI and GDS depressive scores were two variables that emerged as predictors of depression in elderly women classified into the moderate depressive symptoms group. This group exhibited elevated depressive symptoms on the BDI [22] and GDS [25] questionnaires, suggesting moderate depressive symptoms according to their normative data. The elderly women analyzed here had high scores on the BDI questionnaire, mainly for items that evaluated negative thoughts and attitudes, which could contribute to the maintenance of their depressed mood [31]. GDS evaluates depressive symptoms in the elderly people. However, some factors share similarities with BDI, such as sadness, lack of energy, social withdrawal, feelings of worthlessness or despair. In our study, it was notable to observe that BDI and GDS depression scores were higher in the moderate depressive symptoms group. This suggests that the intensity of depressive symptoms was detected in a similar way by both questionnaires. In a sample with similar characteristics, moderate depressive symptoms were also reported in middle-aged women without major depression [5], which is consistent with our results.

In the current research, comorbidities were another variable that arose as a predictor. The comorbidities were higher in the minimal depressive symptoms group and lower in the moderate depressive symptoms group. The most prevalent comorbidities reported by the elderly women included hypertension, osteoporosis, arthritis, gynecological disease, surgery and kidney disease. Other diseases included glaucoma, migraine, hepatitis, hypothyroidism, diabetes, allergies and sinusitis. These comorbidities have been related to depressive symptoms [6,32], which partially coincides with our outcomes. These multiple comorbidities found in elderly women analyzed here makes them delicate individuals. A study conducted in elderly hospitalized individuals demonstrated that malignancy, diabetes, coronary artery disease, chronic kidney disease and chronic obstructive pulmonary disease were more frequent in men, but hypertension, osteoarthritis, anemia and depression were more frequent in women [33]. These findings coincide and support our results, even though we had not considered men in our study, seem to highlight the gender differences that impact the development of diseases in vulnerable elderly. Another study identified different frailty phenotypes, differently associated with adverse events, as multimorbidity and cognitive impairment in old patients [34], which is consistent our outcomes. Some studies have reported certain protective factors associated with depressive symptoms in older adults. Social support to manage health problems and the role engagement in social activities, volunteer work or religion have a significant influence on the manifestation of depression [35]. One study found that the hobbies and indoor activities was associated with lower odds of elevated symptoms for men and women, while the volunteer and community activities was associated with lower odds of depressive symptoms for women [36]. In our study, elderly women attended a senior center that provides recreational and occupational activities, which could influence depression and comorbidities. It has been suggested that resilience in some elderly individuals can support adaptation to comorbidities for successful aging [37], which could explain, at least in part, our findings in the moderate depressive symptoms group.

In the present study, MMSE scores were other variable that emerged as a predictor of moderate depressive symptoms. It has been reported that depressive symptoms predict cognitive decline in elderly individuals [38,39]. Specifically, a decrease in scores for global cognitive function (MMSE) and cognitive flexibility was associated with depressive symptoms in nondemented elderly women [40]. In addition, elderly individuals with severe depression had twice risk for cognitive dysfunction [41]. These results partially support our findings found in the moderate depressive symptoms group. Furthermore, in our research, the elderly women were recruited from a senior center, which provides health care and recreational activities. In this non-clinical sample, no signs of psychosis or other states such as bipolar disorders were detected. Therefore, the results in the MMSE could not be influenced by these disorders. Moreover, it was remarkable to observe that the scores obtained in the MMSE did not show significant differences between the two study groups. However, the discriminant analysis established suitable group membership for MMSE scores.

Notably, in our investigation, L-isoleucine and L-leucine were two essential amino acids that emerged as predictors of moderate depressive symptoms. L-Isoleucine is an essential amino acid that the body cannot manufacture, so it must be obtained through the diet. The benefits of L-isoleucine may include regulating blood glucose, participating in hemoglobin synthesis, and reducing postworkout fatigue and central fatigue [42,43]. L-Leucine is an activator of the mammalian target of rapamycin that regulates protein synthesis, tissue regeneration and metabolism [44]. In the brain, L-leucine serves as a metabolic precursor of fuel molecules that are forwarded by astrocytes to adjacent neural cells for brain energy metabolism [45]. L-Leucine also participates in the regulation of the neurotransmitter glutamate [46]. The moderate depressive symptoms group of elderly women analyzed here had less L-isoleucine and less L-leucine, suggesting deficits in protein synthesis and brain energy metabolism. These deficits also suggest alterations in neurotransmitter glutamate and glucose metabolism, which are crucial to brain function and neural cell vitality [47]. These deficits could explain, at least in part, our outcomes found in elderly women with moderate depressive symptoms. One study revealed that patients diagnosed with bipolar disorder and major depression showed low levels of leucine and isoleucine, and these amino acids were negatively correlated with depressive scores [16,17], which partially supports our findings.

In our investigation, L-histidine was another amino acid that emerged as a predictor of moderate depressive symptoms. L-Histidine is an essential amino acid that has a role in protein components and is a precursor of histamine involved in inflammation. Brain histamine is synthesized from L-histidine in the presence of histidine decarboxylase to play a role as a neurotransmitter in the brain to balance mood, the sleep cycle, learning, memory, alertness, regulation of appetite and the perception of pain [48]. Since histamine is required for many brain processes, low levels of histamine can manifest as depression, anxiety and poor motivation [48], probably related to a reduction in histamine receptors [49]. The moderate depressive symptoms group of elderly women analyzed here had less L-histidine, suggesting deficits in brain histamine, which could induce depressive symptoms. Our outcome is partially supported by a study that showed that histidine intake improved depression, attentiveness, concentration and mental task performance [50].

The findings of the present study have implications for clinical practice in the development of strategies for the prevention of depressive symptoms associated with protein deficiency in vulnerable older adults. Our results will also serve to apply treatments with balanced diets to decrease the manifestation of depressive symptoms in elderly individuals. However, the present study has limitations: First, we did not include elderly women diagnosed with major depression that could help support the effects of essential amino acids in severe depression. Second, we did not evaluate the food intake of elderly women that could help clarify the relationship with protein intake and depressive symptoms. Third, we did not include elderly men without major depression to establish comparisons with gender. Fourth, we did not include elderly women without depression to establish comparisons with the group studied here with the moderate depressive symptoms, which should be included in future studies. The strength of this study is that we used a discriminant analysis to establish deficiencies in essential amino acids related to moderate depressive symptoms, which could reflect alterations in nutrition. Even though the present study did not measure food intake, protein deficiency has been associated with depression in older adults [51] and middle-aged individuals [52]. These results partially support our findings found in elderly women. Since essential amino acids, L-isoleucine and L-leucine, serve as signaling molecules regulating metabolism of glucose, lipid, and protein synthesis, and immunity, a dietary optimizing of these essential amino acids will have health benefits [53].

5. Conclusions

In summary, our results emphasize that the moderate depressive symptoms group was characterized by higher BDI, GDS, and global cognitive scores; fewer comorbidities; and less concentration L-histidine, L-isoleucine and L-leucine. These results suggest that this group showed a deficiency in essential amino acids involved in metabolism, protein synthesis, inflammation and brain neurotransmission, which could influence depressive symptomatology. This insufficiency could be related to nutritional deficiencies, particularly protein intake.

Author Contributions: S.S.-O.: Conceptualization, Funding acquisition, Resources, Methodology, Project administration, Formal analysis, Investigation, Data curation, Data discussion, Supervision, Writing—original draft, Writing—review & editing.; R.G.-G.: Conceptualization, Funding acquisition, Methodology.; V.A.-A.: Methodology (determination of amino acids).; A.T.-B.: Methodology (sample collection). All authors have read and agreed to the published version of the manuscript.

Funding: This study was supported by the Secretaría de Salud del Estado de Guanajuato, México.

Institutional Review Board Statement: The study was conducted according to the guidelines of the Declaration of Helsinki, and approved by the Research and Ethics Committee of the Faculty of Medicine of the National Autonomous University of Mexico (No. 007-2018).

Informed Consent Statement: Informed consent was obtained from all subjects involved in the study.

Acknowledgments: The authors thank Lisette Morado for her participation in the sample collection. This article was written in Silvia Solis-Ortiz's sabbatical year.

Conflicts of Interest: The authors declare no conflict of interest.

References

1. World Health Organization. Depression 2020. Available online: http://who.int/mediacentre/factsheets/fs369/en/ (accessed on 30 January 2020).
2. Agüera-Ortiz, L.; Claver-Martín, M.D.; Franco-Fernández, M.D.; López-Álvarez, J.; Martín-Carrasco, M.; Ramos-García, M.I.; Sánchez-Pérez, M. Depression in the Elderly. Consensus Statement of the Spanish Psychogeriatric Association. *Front. Psychiatry* **2020**, *11*. [CrossRef]
3. Santos, P.V.D.L.; Valdés, S.E.C. Prevalencia de depresión en hombres y mujeres mayores en México y factores de riesgo. *Poblac. Salud Mesoam.* **2017**, *15*. [CrossRef]
4. Gutsmiedl, K.; Krause, M.; Bighelli, I.; Schneider-Thoma, J.; Leucht, S. How well do elderly patients with major depressive disorder respond to antidepressants: A systematic review and single-group meta-analysis. *BMC Psychiatry* **2020**, *20*, 102. [CrossRef]
5. Solís-Ortiz, S.; Pérez-Luque, E.; Pacheco-Zavala, M.D.P. Resting EEG Activity and Ovarian Hormones as Predictors of Depressive Symptoms in Postmenopausal Women without a Diagnosis of Major Depression. *Psychology* **2012**, *03*, 834–840. [CrossRef]
6. Kang, H.-J.; Kim, S.-Y.; Bae, K.-Y.; Kim, S.-W.; Shin, I.-S.; Yoon, J.-S.; Kim, J.-M. Comorbidity of Depression with Physical Disorders: Research and Clinical Implications. *Chonnam Med. J.* **2015**, *51*, 8–18. [CrossRef]
7. National Institute of Aging. Health, Depression and Older Adults. 2020. Available online: www.nih.gob (accessed on 2 March 2020).

8. Hakkarainen, R.; Partonen, T.; Haukka, J.; Virtamo, J.; Albanes, D.; Lönnqvist, J. Association of dietary amino acids with low mood. *Depress. Anxiety* **2003**, *18*, 89–94. [CrossRef]

9. Aucoin, M.; LaChance, L.; Cooley, K.; Kidd, S. Diet and Psychosis: A Scoping Review. *Neuropsychobiology* **2018**, *79*, 20–42. [CrossRef]

10. Loureiro, C.M.; Da Roza, D.L.; Corsi-Zuelli, F.; Shuhama, R.; Fachim, H.A.; Simões-Ambrosio, L.M.C.; Deminice, R.; Jordão, A.A.; Menezes, P.R.; Del-Ben, C.M.; et al. Plasma amino acids profile in first-episode psychosis, unaffected siblings and community-based controls. *Sci. Rep.* **2020**, *10*, 21423. [CrossRef]

11. Leppik, L.; Kriisa, K.; Koido, K.; Koch, K.; Kajalaid, K.; Haring, L.; Vasar, E.; Zilmer, M. Profiling of Amino Acids and Their Derivatives Biogenic Amines Before and After Antipsychotic Treatment in First-Episode Psychosis. *Front. Psychiatry* **2018**, *9*, 155. [CrossRef]

12. Van der Does, A. The effects of tryptophan depletion on mood and psychiatric symptoms. *J. Affect. Disord.* **2001**, *64*, 107–119. [CrossRef]

13. Parker, G.; Brotchie, H. Mood effects of the amino acids tryptophan and tyrosine. *Acta Psychiatr. Scand.* **2011**, *124*, 417–426. [CrossRef]

14. Baranyi, A.; Amouzadeh-Ghadikolai, O.; Von Lewinski, D.; Breitenecker, R.J.; Rothenhäusler, H.-B.; Robier, C.; Baranyi, M.; Theokas, S.; Meinitzer, A. Revisiting the tryptophan-serotonin deficiency and the inflammatory hypotheses of major depression in a biopsychosocial approach. *PeerJ* **2017**, *5*. [CrossRef]

15. Pan, J.-X.; Xia, J.-J.; Deng, F.-L.; Liang, W.-W.; Wu, J.; Yin, B.-M.; Dong, M.-X.; Chen, J.-J.; Ye, F.; Wang, H.-Y.; et al. Diagnosis of major depressive disorder based on changes in multiple plasma neurotransmitters: A targeted metabolomics study. *Transl. Psychiatry* **2018**, *8*, 1–10. [CrossRef]

16. Fellendorf, F.; Platzer, M.; Pilz, R.; Rieger, A.; Kapfhammer, H.-P.; Mangge, H.; Dalkner, N.; Zelzer, S.; Meinitzer, A.; Birner, A.; et al. Branched-chain amino acids are associated with metabolic parameters in bipolar disorder. *World J. Biol. Psychiatry* **2018**, *20*, 821–826. [CrossRef]

17. Baranyi, A.; Amouzadeh-Ghadikolai, O.; von Lewinski, D.; Rothenhäusler, H.B.; Theokas, S.; Robier, C.; Mangge, H.; Reicht, G.; Hlade, P.; Meinitzer, A. Branched-Chain amino acids as new biomarkers of major depression-A novel neurobiology of mood disorder. *PLoS ONE* **2016**, *11*, e0160542. [CrossRef]

18. Carvalho, A.F.; Solmi, M.; Sanches, M.; Machado, M.O.; Stubbs, B.; Ajnakina, O.; Sherman, C.; Sun, Y.R.; Liu, C.S.; Brunoni, A.R.; et al. Evidence-based umbrella review of 162 peripheral biomarkers for major mental disorders. *Transl. Psychiatry* **2020**, *10*, 1–13. [CrossRef]

19. Fisher, R.A. The use of multiple measurements in taxonomic problems. *Ann. Eugen.* **1936**, *7*, 179–188. [CrossRef]

20. Flury, B.; Reidwyl, H. Discriminant analysis. In *Multivariate Statistics. A Practical Approach*; Chapman and Hall: London, UK, 1988; pp. 181–233.

21. American Psychiatric Association. *Diagnostic and Statistical Manual of Mental Disorders*, 5th ed.; American Psychiatric Association: Arlington, VA, USA, 2013.

22. Beck, A.T.; Steer, R.A. *Beck Depression Inventory*; The Psychological Corporation: San Antonio, TX, USA, 1993.

23. Kuehner, C. Why is depression more common among women than among men? *Lancet Psychiatry* **2017**, *4*, 146–158. [CrossRef]

24. Jurado, S.; Villegas, M.E.; Méndez, L.; Rodríguez, F.; Loperena, V.; Varela, R. La estandarización del inventario de la depresión de Beck, para los residentes de la ciudad de México. *Salud Ment.* **1998**, *21*, 36–38.

25. Yesavage, J.A.; Brink, T.; Rose, T.L.; Lum, O.; Huang, V.; Adey, M.; Leirer, V.O. Development and validation of a geriatric depression screening scale: A preliminary report. *J. Psychiatr. Res.* **1983**, *17*, 37–49. [CrossRef]

26. Martínez-de la Iglesia, J.; Onís-Vilches, M.C.; Dueñas-Herrero, R.; Albert-Colomer, C.; Aguado-Taberné, C.; Luque, D. Versión española del cuestionario de Yesavage abreviado (GDS) para el despistaje de depresión en mayores de 65 años: Adaptación y validación. *MEDIFAM* **2002**, *12*, 620–630. [CrossRef]

27. Folstein, M.F.; Folstein, S.E.; McHugh, P.R. "Mini-mental state": A practical method for grading the cognitive state of patients for the clinician. *J. Psychiatr. Res.* **1975**, *12*, 189–198. [CrossRef]

28. Lobo, A.; Saz, P.; Marcos, G.; Grupo de Trabajo ZARADEMP. *MMSE: Examen Cognoscitivo Mini-Mental*; TEA Ediciones: Madrid, Spain, 2002.

29. Henderson, J.W.; Ricker, R.D.; Bidlingmeyer, B.A.; Woodward, C. Rapid, accurate, sensitive and reproducible HPLC analysis of amino acids. In *Amino Acid Analysis using Zorbax Aclipse AAA Columns and the Agilent 1100 HPLC*; Agilent Technologies: Santa Clara, CA, USA, 2000.

30. Field, A. Multivariate Analysis of Variance. In *Discovering Statistics Using SPSS*; Sage Publications Ltd.: London, UK, 2009; pp. 587–623.

31. Dowd, E.T. Depression: Theory, assessment, and new directions in practice. *Int. J. Clin. Health Psychol.* **2004**, *4*, 413–423.

32. Steffen, A.; Nübel, J.; Jacobi, F.; Bätzing, J.; Holstiege, J. Mental and somatic comorbidity of depression: A comprehensive cross-sectional analysis of 202 diagnosis groups using German nationwide ambulatory claims data. *BMC Psychiatry* **2020**, *20*, 142. [CrossRef]

33. Corrao, S.; Santalucia, P.; Argano, C.; Djade, C.; Barone, E.; Tettamanti, M.; Pasina, L.; Franchi, C.; Eldin, T.K.; Marengoni, A.; et al. Gender-differences in disease distribution and outcome in hospitalized elderly: Data from the REPOSI study. *Eur. J. Intern. Med.* **2014**, *25*, 617–623. [CrossRef]

34. Marcucci, M.; Franchi, C.; Nobili, A.; Mannucci, P.M.; Ardoino, I. REPOSI Investigators Defining Aging Phenotypes and Related Outcomes: Clues to Recognize Frailty in Hospitalized Older Patients. *J. Gerontol. Ser. A Boil. Sci. Med Sci.* **2016**, *72*, 395–402. [CrossRef]

35. Fiske, A.; Wetherell, J.L.; Gatz, M. Depression in Older Adults. *Annu. Rev. Clin. Psychol.* **2009**, *5*, 363–389. [CrossRef] [PubMed]

36. Gutierrez, S.; Milani, S.A.; Wong, R. Is "Busy" Always Better? Time-Use Activities and Depressive Symptoms Among Older Mexican Adults. *Innov. Aging* **2020**, *4*, igaa030. [CrossRef]

37. Pruchno, R.; Carr, D. Successful Aging 2.0: Resilience and Beyond. *J. Gerontol. Ser. B* **2017**, *72*, 201–203. [CrossRef]

38. Wilson, R.S.; Mendes, D.; Bennett, D.; Bienias, J.L.; Evans, D. Depressive symptoms and cognitive decline in a community population of older persons. *J. Neurol. Neurosurg. Psychiatry* **2004**, *75*, 126–129.

39. Perini, G.; Ramusino, M.C.; Sinforiani, E.; Bernini, S.; Petrachi, R.; Costa, A. Cognitive impairment in depression: Recent advances and novel treatments. *Neuropsychiatr. Dis. Treat.* **2019**, *15*, 1249–1258. [CrossRef]

40. Yaffe, K.; Blackwell, T.; Gore, R.; Sands, L.; Reus, V.; Browner, W.S. Depressive Symptoms and Cognitive Decline in Nondemented Elderly Women. *Arch. Gen. Psychiatry* **1999**, *56*, 425–430. [CrossRef] [PubMed]

41. Aajami, Z.; Kazazi, L.; Troski, M.; Bahrami, M.; Borhaninejad, V. Relationship between Depression and Cognitive Impairment among Elderly: A Cross-sectional Study. *J. Caring Sci.* **2020**, *9*, 148–153. [CrossRef]

42. HoleČek, M. Branched-chain amino acids in health and disease: Metabolism, alterations in blood plasma, and as supplements. *Nutr. Metab.* **2018**, *15*, 1–12. [CrossRef]

43. Blomstrand, E. A Role for Branched-Chain Amino Acids in Reducing Central Fatigue. *J. Nutr.* **2006**, *136*, 544S–547S. [CrossRef]

44. Pedroso, J.A.; Zampieri, T.T.; Donato, J.J. Reviewing the Effects of L-Leucine Supplementation in the Regulation of Food Intake, Energy Balance, and Glucose Homeostasis. *Nutrients* **2015**, *7*, 3914–3937. [CrossRef]

45. Murín, R.; Hamprecht, B. Metabolic and Regulatory Roles of Leucine in Neural Cells. *Neurochem. Res.* **2007**, *33*, 279–284. [CrossRef]

46. García-Espinosa, M.A.; Wallin, R.; Hutson, S.M.; Sweatt, A.J. Widespread neuronal expression of branched-chain aminotransferase in the CNS: Implications for leucine/glutamate metabolism and for signaling by amino acids. *J. Neurochem.* **2007**, *100*, 1458–1468. [CrossRef]

47. Sperringer, J.E.; Addington, A.; Hutson, S.M. Branched-Chain Amino Acids and Brain Metabolism. *Neurochem. Res.* **2017**, *42*, 1697–1709. [CrossRef]

48. Haas, H.L.; Sergeeva, O.A.; Selbach, O. Histamine in the Nervous System. *Physiol. Rev.* **2008**, *88*, 1183–1241. [CrossRef]

49. Kano, M.; Fukudo, S.; Tashiro, A.; Utsumi, A.; Tamura, D.; Itoh, M.; Iwata, R.; Tashiro, M.; Mochizuki, H.; Funaki, Y.; et al. Decreased histamine H1 receptor binding in the brain of depressed patients. *Eur. J. Neurosci.* **2004**, *20*, 803–810. [CrossRef]

50. Sasahara, I.; Fujimura, N.; Nozawa, Y.; Furuhata, Y.; Sato, H. The effect of histidine on mental fatigue and cognitive performance in subjects with high fatigue and sleep disruption scores. *Physiol. Behav.* **2015**, *147*, 238–244. [CrossRef] [PubMed]

51. Park, Y.-H.; Choi-Kwon, S.; Park, K.-A.; Suh, M.; Jung, Y. Nutrient deficiencies and depression in older adults according to sex: A cross sectional study. *Nurs. Health Sci.* **2016**, *19*, 88–94. [CrossRef] [PubMed]

52. Oh, J.; Yun, K.; Chae, J.-H.; Kim, T.-S. Association between Macronutrients Intake and Depression in the United States and South Korea. *Front. Psychiatry* **2020**, *11*, 207. [CrossRef]

53. Nie, C.; He, T.; Zhang, W.; Zhang, G.; Ma, X. Branched Chain Amino Acids: Beyond Nutrition Metabolism. *Int. J. Mol. Sci.* **2018**, *19*, 954. [CrossRef]

Article

The Inverse Correlation of Isoflavone Dietary Intake and Headache in Peri- and Post-Menopausal Women

Mayuko Kazama [1], Masakazu Terauchi [2,*], Tamami Odai [2], Kiyoko Kato [2] and Naoyuki Miyasaka [1]

[1] Department of Obstetrics and Gynecology, Tokyo Medical and Dental University, Tokyo 113-8510, Japan; 180202ms@tmd.ac.jp (M.K.); n.miyasaka.gyne@tmd.ac.jp (N.M.)

[2] Department of Women's Health, Tokyo Medical and Dental University, Tokyo 113-8510, Japan; odycrm@tmd.ac.jp (T.O.); kiyo.crm@tmd.ac.jp (K.K.)

* Correspondence: teragyne@tmd.ac.jp

Abstract: This study investigated the relationship between headache and dietary consumption of a variety of nutrients in middle-aged women. This cross-sectional analysis used first-visit records of 405 women aged 40–59 years. The frequency of headaches was assessed using the Menopausal Health-Related Quality of Life Questionnaire. Of the 43 major nutrient intakes surveyed using the brief-type self-administered diet history questionnaire, those that were not shared between women with and without frequent headaches were selected. Multiple logistic regression analysis was used to identify nutrients independently associated with frequent headaches. After adjusting for background factors related to frequent headache (vasomotor, insomnia, anxiety, and depression symptoms), the estimated dietary intake of isoflavones (daidzein + genistein) (mg/1000 kcal/day) was negatively associated with frequent headaches (adjusted odds, 0.974; 95% confidence interval, 0.950–0.999). Moreover, the estimated isoflavone intake was not significantly associated with headache frequency in the premenopausal group, whereas it significantly correlated with that in the peri- and post-menopausal groups. Headache in peri- and post-menopausal women was inversely correlated with the dietary intake of isoflavones. Diets rich in isoflavones may improve headaches in middle-aged women.

Keywords: menopause; isoflavone; daidzein; genistein; headache

Citation: Kazama, M.; Terauchi, M.; Odai, T.; Kato, K.; Miyasaka, N. The Inverse Correlation of Isoflavone Dietary Intake and Headache in Peri- and Post-Menopausal Women. *Nutrients* **2022**, *14*, 1226. https://doi.org/10.3390/nu14061226

Academic Editor: LaVerne L. Brown

Received: 14 January 2022
Accepted: 8 March 2022
Published: 14 March 2022

Publisher's Note: MDPI stays neutral with regard to jurisdictional claims in published maps and institutional affiliations.

Copyright: © 2022 by the authors. Licensee MDPI, Basel, Switzerland. This article is an open access article distributed under the terms and conditions of the Creative Commons Attribution (CC BY) license (https://creativecommons.org/licenses/by/4.0/).

1. Introduction

Headache is quite common among women and is one of the most prevalent symptoms of menopause [1]. Most menopausal symptoms checklists include headache, and in our previous cross-sectional study, headaches were the 13th most common menopausal symptom of the 21 categories, with 52.8% of middle-aged women complaining of headache at least once a week [2]. The two most common subtypes of headache are tension headaches and migraines. The former is reported to be unchanged or worsened after menopause in more than two-thirds of female patients [3]. Contrarily, as the latter can be triggered by estrogen withdrawal, symptoms tend to improve after menopause [4] although it is still reported in 11–24% of post-menopausal women [4].

Studies have shown that specific diets, such as a low-calorie diet, carbohydrate-restricted diet, and weight-loss diets for obese patients, can improve headaches [5]. In a randomized controlled trial of 182 patients, high n-3 fatty acid diets ameliorated the frequency and severity of headache compared to the control diet [6].

The effect of diet on fluctuating estrogen-related headaches has also been studied. Namely, vitamin E intake has been reported to improve menstrual migraine in a double-blind controlled trial [1]. However, there have been few reports on the relationship between diet or nutrients and headaches especially in women transitioning through menopause. This study aimed to investigate the relationship between headache and dietary nutrient intake in middle-aged Japanese women.

2. Materials and Methods

2.1. Study Population

In the present cross-sectional study, we used the first-visit records of 2090 Japanese women enrolled in the Systematic Health and Nutrition Education Program (SHNEP) conducted at the Menopause Clinic of Tokyo Medical and Dental University, Tokyo, Japan, between March 1995 and January 2021. All participants in the program visited or were referred to our clinic for the treatment of menopausal symptoms. Of these, 494 responded to the Menopausal Health-Related Quality of Life Questionnaire (MHR-QOL) [7] and the brief-type self-administered dietary history questionnaire (BDHQ) [8,9]. The 46 patients treated with hormone replacement therapy, 35 patients younger than 40 years or older than 59 years, and 4 patients with unknown menopausal status were excluded, leaving 409 patient records included for analysis.

The research protocol was reviewed and approved by the Tokyo Medical and Dental University Review Board (number: 774, approval date: 23 March 2010), and written informed consent was obtained from all participants. This study was conducted following the Declaration of Helsinki.

2.2. Menopausal Status

We classified the women into three menopausal statuses: premenopausal, perimenopausal, and postmenopausal. The premenopausal group had regular menstrual periods over the past three months, the perimenopausal group had experienced menstruation in the past 12 months but not in the past three months or had irregular menstrual cycles, and the postmenopausal group had no menstruation in the past 12 months.

2.3. Physical Assessment

Body composition variables, such as body mass index, body fat percentage, fat mass, lean mass, muscle mass, and basal metabolism, were measured using a bioimpedance analyzer (MC190-EM; Tanita, Tokyo, Japan). Height, weight, and waist and hip circumference were measured to calculate body mass index and the waist–hip ratio. Body temperature was measured using a thermometer. Resting metabolic rate was estimated from the respiratory volume using an indirect calorimeter (Metavine-N VMB-005 N; Vine, Tokyo, Japan). Additionally, cardiovascular parameters, including blood pressure, heart rate, cardio-ankle vascular index, and ankle-brachial pressure index, were measured using a vascular screening system (VS-1000; Fukuda Denshi, Tokyo, Japan). We also conducted a physical fitness test to assess power, reaction time, and flexibility. Hand-grip strength was measured twice with a hand dynamometer (Yagami, Nagoya, Japan), and the mean value (kgf) was calculated using the larger value of the two measurements. The test for reaction time was repeated seven times with the ruler-drop test using a wooden ruler (Yagami, Nagoya, Japan) with a length of 60 cm and a weight of 110 g. The ruler-drop test was carried out by allowing a seated participant to affix their arm on a desk, extending their hand from the edge, while the examiner holds out the ruler above the participant's thumb and index finger. The participant then attempts to catch the ruler as quickly as possible when it drops. The locations a participant grasped the ruler for each trial were evaluated, and the average reaction time (cm) was calculated from the remaining five values omitting the maximum and minimum values. Flexibility was measured by forward bend test (cm) using a reach box while sitting (Yagami, Nagoya, Japan).

2.4. Lifestyle Characteristics

We surveyed lifestyle factors, including smoking history (none, less than 20 cigarettes per day, more than 20 cigarettes per day), alcohol consumption (never, sometimes, daily), caffeine consumption (never, 1–2 times per day, 3 or more times per day), and regular exercise habits (yes, no).

2.5. Questionnaires

The MHR-QOL, developed and validated in our clinic [7], is a modification of the Women's Health Questionnaire and other questionnaires [10–12]. Physical and psychological symptoms were scored on a 4-point Likert scale based on the frequency of symptoms. In this analysis, the scores increased as each symptom became more frequent (0, zero to once a month; 1, one-two times per week; 2, three-four times per week; 3, almost every day). In the present study, the participants that answered "3" under headache were defined as having frequent headaches, while all others were defined as without frequent headaches. The sum of the scores under hot flashes and night sweats was used as the vasomotor symptom (VMS) score (0–6). Likewise, the scores under difficulty in initiating sleep and non-restorative sleep were pooled to define the insomnia symptom score (0–6).

The Hospital Anxiety and Depression Scale (HADS) is a widely used and reliable screening test for mental health in patients with somatic symptoms [13]. The HADS consists of seven items for two subscales, depression and anxiety symptoms, and participants respond to each item on a 4-point Likert scale. Patients who rated 8–10 were classified as likely to have anxiety or depression, while those who rated 11–21 were classified as having anxiety or depression.

The BDHQ is a self-administered questionnaire that assesses the participant's intake of 61 food items commonly consumed in Japan. The participants answered 77 questions regarding their consumption over the previous month. Based on their responses, the intakes of 96 nutrients were estimated using an ad hoc computer algorithm. A semi-weight method was used to adjust for total calorie intake [8,9]. In this study, we investigated the association between the frequency of headaches and the estimated intake of 43 major nutrients.

2.6. Statistical Analysis

Continuous variables are presented as mean $\pm$ standard deviation (SD). The required sample size, determined to be 333, was derived by multiplying 10 with the number of independent variables, then dividing the product by the prevalence of frequent headaches (estimated to be 5 and 0.15, respectively). Differences between the groups were compared using an unpaired *t*-test, Mann–Whitney test, chi-square test, and Fisher's exact test. Multicollinearity between variables was determined using the cutoff points for Pearson or Spearman correlation coefficients of $|R| > 0.9$. The nutrients and socio-demographic factors that significantly differed between the group with and without frequent headaches were selected. After adjustments for the selected socio-demographic factors, the association between the frequency of headaches and the selected nutrients was analyzed using a multivariate logistic regression model. Statistical significance was set at $p < 0.05$. All Statistical analyses were performed with GraphPad Prism version 9.1.2 (GraphPad Software, San Diego, CA, USA) and JMP version 14.0.2 (SAS Institute Inc., Cary, NC, USA).

3. Results

The participants' average age was 50.1 ± 3.8 years (mean $\pm$ SD). The number (percentage) of women who scored the frequency of their headaches as 0 was 176 (43.0%), while 122 (29.8%) answered 1, 51 (12.5%) answered 2, and 60 (14.7%) answered 3, indicating that more than half of the participants suffered from a headache at least once a week. The socio-demographic factors that were significantly different between the two groups, with and without frequent headaches, are presented in Table 1. Participants with frequent headaches had higher VMS, insomnia, anxiety, and depression scores than those without. We also assessed the dietary intake of 43 major nutrients, which differed significantly between the two groups (Table 2). The estimated intakes of vitamin K, daidzein, and genistein were lower in women with frequent headaches than in those without. A test of multicollinearity for these variables showed that the correlation coefficient between daidzein and genistein was 1.00. Therefore, the estimated isoflavone intake was defined as the sum of daidzein and genistein. The values of the other correlation coefficients were less than 0.7.

Table 1. Comparison of socio-demographic factors and the frequency of headache.

Characteristic	Headache		*p*-Value
	Frequent (*N* = 60)	Not So Frequent (*N* = 349)	
Age and Menopausal Status			
Age (years), mean (SD)	49.4 (3.9)	50.2 (3.8)	0.128 [a]
Menopausal status, premenopausal/peri-menopausal/postmenopausal (%)	37.3/27.1/35.6	32.4/23.1/44.5	0.441 [b]
Body composition, mean (SD)			
Height (cm)	157.4 (5.0)	157.6 (5.4)	0.839 [c]
Weight (kg)	54.0 (10.8)	54.4 (9.6)	0.691 [a]
Body mass index (kg/m^2)	21.8 (4.1)	21.9 (3.8)	0.616 [a]
Waist–hip ratio	0.87 (0.07)	0.87 (0.06)	0.762 [c]
Fat mass (kg)	15.9 (8.2)	15.8 (7.1)	0.899 [a]
Lean mass (kg)	38.1 (3.3)	38.5 (3.5)	0.355 [c]
Muscle mass (kg)	35.9 (3.1)	36.3 (3.2)	0.355 [c]
Body fat percentage (%)	27.9 (8.8)	28.0 (7.7)	0.927 [c]
Basal metabolism (MJ/day)	1107 (125)	1118 (122)	0.491 [a]
Resting energy expenditure (MJ/day)	1664 (400)	1578 (428)	0.171 [a]
Body temperature (°C)	36.2 (0.5)	36.3 (0.5)	0.765 [a]
Physical fitness test, mean (SD)			
Hand-grip strength (kg)	25.2 (4.8)	25.3 (4.8)	0.834 [c]
Ruler-drop test (cm)	22.9 (4.8)	22.8 (4.0)	0.895 [a]
Sit-and-reach test (cm)	35.9 (10.0)	36.1 (9.7)	0.794 [a]
Cardiovascular parameters, mean (SD)			
Systolic blood pressure (mmHg)	128 (16)	126.5 (17)	0.433 [a]
Diastolic blood pressure (mmHg)	82.0 (12.5)	80.8 (12.2)	0.350 [a]
Heart rate (per min)	65.0 (11.3)	63.7 (11.7)	0.441 [a]
Cardio-ankle vascular index	7.47 (1.12)	7.42 (0.69)	0.148 [a]
Ankle-brachial pressure index	1.11 (0.08)	1.10 (0.06)	0.077 [c]
Lifestyle factors (%)			
Smoking (cigarettes per day)			
None/fewer than 20/20 or more	88.3/8.3/3.3	89.7/7.5/2.9	0.953 [b]
Drinking			
Never/sometimes/daily	63.3/25.0/11.7	56.9/31.0/12.1	0.609 [b]
Caffeine (per day)			
Never/1–2 times/3 or more times	13.3/36.7/50.0	6.9/34.5/58.6	0.179 [b]
Exercise			
Moderate exercise Yes/no	51.7/48.3	43.0/57.0	0.260 [d]
Regular exercise Yes/no	40.0/60.0	40.3/60.0	>0.999 [d]
Physical symptoms, mean (SD)			
MHR-QOL vasomotor symptom score (0–6)	3.4 (2.3)	2.3 (2.1)	0.001 [a]
Psychological symptoms, mean (SD)			
MHR-QOL insomnia symptom score (0–6)	3.7 (2.2)	2.3 (2.1)	<0.001 [a]
Hospital Anxiety and Depression scale Anxiety subscale score (0–21)	9.2 (4.3)	7.8 (3.7)	0.029 [a]
Hospital Anxiety and Depression scale Depression subscale score (0–21)	8.7 (4.4)	7.3 (3.7)	0.024 [a]

[a] Mann–Whitney test. [b] Chi-square test. [c] Unpaired *t*-test. [d] Fisher's exact test. *N*, number; SD, standard deviation; MHR-QOL: Menopausal Health-Related Quality of Life Questionnaire.

Afterwards, multivariate logistic regression analysis was performed to reveal an independent association between the intake of selected nutrients (vitamin K and isoflavones) and the frequency of headache (Table 3). After adjustment for the selected nutrients (Model 1) and the socio-demographic factors significantly related to the frequency of headache (Model 2), the estimated intake of isoflavone remained significantly associated with frequent headache (Model 1: adjusted odds ratio (OR) per mg/1000 kcal/day = 0.976, 95% confidence interval (CI) = 0.952–1.000, *p* = 0.048; Model 2: OR = 0.974, CI = 0.950–0.999, *p* = 0.036), while the intake of vitamin K was not. In Model 2, VMS and insomnia scores had a significant relationship with the frequency of headache (VMS: OR = 1.160, CI = 1.012–1.329, p = 0.033; insomnia: OR = 1.334, CI = 1.152–1.546, p < 0.001).

Table 2. Comparison of the daily intake of nutrients and the frequency of headache.

Characteristic	Headache		p-Value
	Frequent (N = 60)	Not So Frequent (N = 349)	
Nutrition, Mean (SD)			
Protein (%E)	15.5 (3.3)	15.8 (3.1)	0.244 [a]
Animal protein (%E)	8.8 (3.3)	9.0 (3.1)	0.585 [a]
Vegetable protein (%E)	6.7 (1.3)	6.8 (1.2)	0.511 [b]
Carbohydrate (%E)	53.5 (10.1)	52.2 (8.6)	0.235 [a]
Ash (g/1000 kcal/day)	10.7 (2.5)	10.6 (2.0)	0.815 [a]
Sodium (mg/1000 kcal/day)	2405 (551)	2316 (478)	0.173 [a]
Potassium (mg/1000 kcal/day)	1525 (588)	1594 (439)	0.054 [a]
Calcium (mg/1000 kcal/day)	325 (127)	336 (112)	0.142 [a]
Magnesium (mg/1000 kcal/day)	147 (41)	153 (35)	0.072 [a]
Phosphorus (mg/1000 kcal/day)	589 (133)	608 (129)	0.174 [a]
Iron (mg/1000 kcal/day)	4.6 (1.5)	4.7 (1.1)	0.218 [a]
Zinc (mg/1000 kcal/day)	4.53 (0.80)	4.58 (0.74)	0.432 [a]
Copper (mg/1000 kcal/day)	0.65 (0.14)	0.66 (0.12)	0.344 [a]
Manganese (mg/1000 kcal/day)	1.79 (0.64)	1.78 (0.61)	0.909 [a]
Fat (%E)	27.2 (6.1)	28.0 (5.8)	0.330 [b]
Animal fat (%E)	12.4 (4.5)	12.8 (4.2)	0.496 [a]
Vegetable fat (%E)	14.8 (4.0)	15.2 (4.0)	0.500 [b]
Saturated fatty acid (%E)	7.4 (2.1)	7.7 (1.9)	0.095 [a]
Monounsaturated fatty acid (%E)	9.6 (2.4)	9.9 (2.3)	0.224 [b]
Polyunsaturated fatty acid (%E)	6.6 (1.7)	6.7 (1.6)	0.350 [a]
Cholesterol (mg/1000 kcal/day)	201 (74)	208 (73)	0.320 [a]
N-3 fatty acid (%E)	1.3 (0.5)	1.4 (0.4)	0.295 [a]
N-6 fatty acid (%E)	5.2 (1.4)	5.3 (1.3)	0.399 [a]
Soluble dietary fiber (g/1000 kcal/day)	1.86 (0.81)	1.99 (0.66)	0.054 [a]
Insoluble dietary fiber (g/1000 kcal/day)	5.18 (2.46)	5.38 (1.65)	0.064 [a]
Dietary fiber (g/1000 kcal/day)	7.31 (3.56)	7.59 (2.37)	0.065 [a]
Daidzein (mg/1000 kcal/day)	7.7 (5.8)	9.2 (5.4)	0.009 [a]
Genistein (mg/1000 kcal/day)	13.0 (9.9)	15.6 (9.1)	0.009 [a]
Isoflavone (mg/1000 kcal/day)	20.7 (15.7)	24.8 (14.5)	0.009 [a]
Retinol (μg/1000 kcal/day)	254 (171)	223 (147)	0.360 [a]
β-carotene (μg/1000 kcal/day)	2495 (2492)	2459 (1522)	0.108 [a]
Retinol equivalent (μg/1000 kcal/day)	463 (262)	430 (202)	0.779 [a]
Vitamin D (μg/1000 kcal/day)	6.92 (4.87)	7.29 (4.75)	0.294 [a]
α-tocopherol (mg/1000 kcal/day)	4.45 (1.60)	4.42 (1.15)	0.417 [a]
Vitamin K (μg/1000 kcal/day)	198 (149)	209 (101)	0.044 [a]
Vitamin B1 (mg/1000 kcal/day)	0.45 (0.13)	0.46 (0.10)	0.183 [a]
Vitamin B2 (mg/1000 kcal/day)	0.77 (0.23)	0.78 (0.20)	0.256 [a]
Niacin (mgNE/1000 kcal/day)	9.64 (3.06)	9.85 (2.68)	0.425 [a]
Vitamin B6 (mg/1000 kcal/day)	0.72 (0.25)	0.75 (0.20)	0.068 [a]
Vitamin B12 (μg/1000 kcal/day)	5.14 (3.06)	5.06 (2.57)	0.677 [a]
Folic acid (μg/1000 kcal/day)	210 (104)	210 (76)	0.356 [a]
Pantothenic acid (mg/1000 kcal/day)	3.67 (0.84)	3.82 (0.77)	0.081 [b]
Vitamin C (mg/1000 kcal/day)	73.6 (41.7)	73.8 (34.0)	0.499 [a]

[a] Mann–Whitney test. [b] Unpaired t-test. %E, % energy.

Table 3. Associations between the daily intake of isoflavone (mg/1000 kcal/day) and the frequency of headache.

	Nutrient	OR	95% CI	*p*-Value
Model 1	Vitamin K (µg/1000 kcal/day)	1.000	0.998–1.004	0.547
	Isoflavone (mg/1000 kcal/day)	0.976	0.952–1.000	0.048
Model 2	Vitamin K (µg/1000 kcal/day)	1.000	0.998–1.004	0.625
	Isoflavone (mg/1000 kcal/day)	0.974	0.950–0.999	0.036
	MHR-QOL VMS score (0–6)	1.160	1.012–1.329	0.033
	MHR-QOL insomnia score (0–6)	1.334	1.152–1.546	<0.001
	HADS depression subscale score (0–21)	1.067	0.967–1.177	0.199
	HADS anxiety subscale score (0–21)	0.994	0.897–1.101	0.902

OR, odds ratio; CI, confidence interval; MHR-QOL, Menopausal Health-Related Quality of Life Questionnaire; VMS, vasomotor symptom; HADS, Hospital Anxiety and Depression Scale. Model 1: Association between daily intake of isoflavone (mg/1000 kcal/day), vitamin K, and headache. Model 2: Multivariate logistic regression model adjusted for VMS, insomnia, depression, and anxiety scores.

Furthermore, in order to investigate the effect of estrogen fluctuation, the participants were divided into two groups: premenopausal (N = 134) and peri- and post-menopausal (N = 271), and the relationship between the frequency of headache and the dietary intake of isoflavone was evaluated in each group (Figure 1). The premenopausal group showed no significant difference in the isoflavone intake (mg/1000 kcal/day) between the women with frequent headaches and those without (21.6 (SD, 16.2) vs. 21.3 (SD, 11.2), p = 0.391, Mann–Whitney test). Contrary, in the peri- and post-menopausal group, the participants suffering from frequent headaches were found to consume significantly less isoflavone than those without (20.3 (SD, 15.8) vs. 26.6 (SD, 15.6), p = 0.011, Mann–Whitney test).

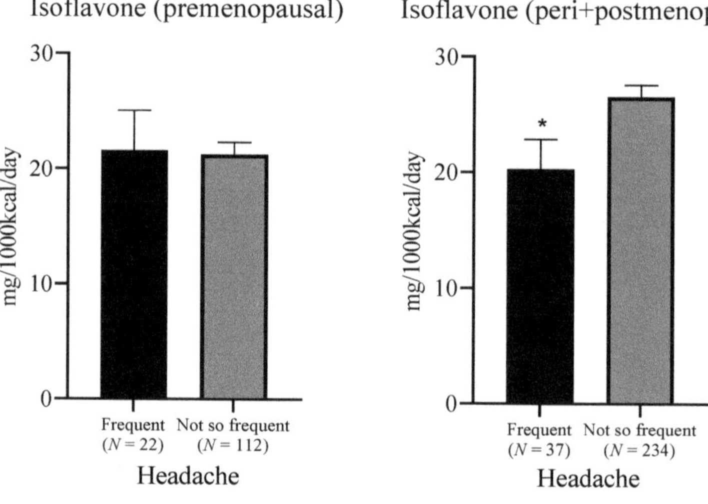

Figure 1. Comparison of daily isoflavone intake by menopausal status between women with frequent headaches and those without. * p < 0.05 vs. Not so frequent, Mann-Whitney test.

4. Discussion

In this cross-sectional study, the estimated dietary intake of isoflavone was inversely associated with frequent headaches in middle-aged Japanese women independent of VMS and insomnia symptom scores. The socio-demographic factors were found to be associated with headaches. A significant difference in isoflavone intake between those with and without headaches was observed in the peri- and post-menopausal groups but not in the premenopausal group.

Isoflavones, including daidzein and genistein, are flavonoids that are especially abundant in leguminous plants. Isoflavones are classified as phytoestrogens because they are structurally similar to 17β-estradiol and have estrogenic and anti-estrogenic effects as well as antioxidant effects [14,15]. Isoflavones have been shown to improve various menopausal symptoms, presumably through an estrogenic effect [16,17]. Isoflavones bind to both estrogen receptors (ER), ERα and ERβ, with a higher affinity for the ERβ receptor [14]. Although the binding activity of isoflavones to the ERs is much weaker than that of estradiol [18], the concentration of serum isoflavone in people consuming soy foods is several orders of magnitude higher than the physiological concentration of estrogen [19], suggesting that isoflavones have a certain physiological effect despite their relatively low ER affinity.

Few reports inferred that isoflavone alleviates premenstrual headache syndrome [20]. In a randomized controlled trial (RCT) of 23 women, consumption of isolated soy protein containing isoflavone (68 mg/day) significantly reduced premenstrual headache [21]. However, to the best of our knowledge, no study has investigated the association between isoflavone intake and headaches in middle-aged women despite headaches being prevalent in women transitioning through menopause (10–29%) [4]. The results of the present study suggest that dietary intake of isoflavone could improve the headache experienced by middle-aged women.

Several pathways could be involved in the association between isoflavone intake and improvement in headache (Figure 2). It has been established that soy isoflavone relieves VMS, especially hot flashes [17,22,23]. The link between VMS and headaches is not clear, but the relationship between VMS and insomnia symptoms is evident [24]. Insomnia symptoms, along with depressive symptoms, improved with a low dose of isoflavone (25 mg/day) in our RCT that enrolled middle-aged women [25]. It is well-known that headaches and insomnia have a bidirectional relationship; headaches can contribute to sleep disturbances, while sleep disturbances can be a predictor or trigger for headache attacks [26]. The other pathways involve anxiety and depression symptoms, known to be associated with headaches. Patients with generalized anxiety disorder (GAD) were at a higher risk for migraine (OR 3.86, 95% CI 2.48 to 6) than those without; accordingly, patients with migraines were at a higher risk for GAD (OR 3.13, 95% CI 1.56 6.3) [27]. Likewise, depression in migraineurs is 2.5 times more common than that in non-migraineurs and is often associated with anxiety symptoms [27]. Several reports, including our recent RCT [25], have shown that isoflavone intake alleviated depressive symptoms in peri- and post-menopausal women [28]. These findings suggest that the relationship between headaches and isoflavone intake could be influenced by the pathways involving VMS and insomnia, depression, and anxiety symptoms. However, in this study, we have shown that isoflavone intake is related to headache independent from these symptoms, which suggests that isoflavone might have a direct effect on headache.

Concerning the mechanism by which isoflavone directly alleviates headache, both its antioxidant and estrogenic effects could be candidates. Considering that the estimated amount of isoflavone intake was different between those with and without headache in the peri- and post-menopausal groups but not in the premenopausal group, the plausible mechanism by which isoflavone reduces headache would be through its estrogenic effect rather than its antioxidant effect. Transitioning through menopause, serum estradiol levels in women fluctuate widely and then decrease [29], which is a trigger of migraine in peri- and post-menopausal women [3]. The estrogenic effect of isoflavone may compensate for the fluctuations and decrease of estrogen [30,31]. Estrogen is known for its regulation of neuronal excitability, interaction with the vascular endothelium in the brain, and association with neurotransmitters, such as serotonin and norepinephrine. [32]. In the cardiovascular and central nervous system, ERβ is predominant, to which isoflavones have a higher affinity [33]. These findings suggest that isoflavone could improve headaches in middle-aged women through its estrogenic effect.

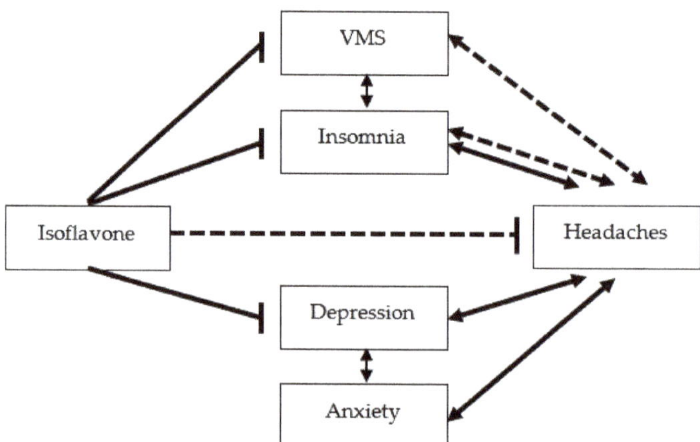

Figure 2. Possible pathways from isoflavone to headache. Arrows, bidirectional relationships; T-bars, inhibitory relationships; solid lines, relationships suggested from previous studies; dashed lines, relationships suggested in this study.

One of the major limitations of this study is the relatively narrow population of research participants attending menopause clinics in Japan. Since the intake of soy, a rich source of isoflavone, varies widely from country to country [34], it would be difficult to generalize this results to a broader population. Additionally, the questionnaire only obtained the frequency of headaches without distinction between different types of headaches, such as migraine and tension headache. Since the pathogeneses of these headaches are different, the relationship between isoflavone intake and each subtype needs to be further investigated although there is a challenge in diagnosing the exact type of headache. Another limitation was the cross-sectional design of our current study, which did not allow us to determine the causal relationship between the estimated intake of isoflavone and headache.

Nevertheless, this study has several strengths. The extensive socio-demographic factors related to headache were simultaneously investigated, including age, menopausal status, body composition, cardiovascular parameters, basal metabolism, physical fitness, lifestyle factors, vasomotor symptoms, and psychological symptoms, in addition to the estimated dietary intake of 43 major nutrients evaluated. As far as we know, this is the first report to show an association between dietary isoflavone intake and menopausal headache symptoms.

5. Conclusions

Headache in peri- and post-menopausal women was inversely associated with dietary intake of isoflavone. Diets rich in isoflavones might improve headaches in middle-aged women.

Author Contributions: M.K., M.T. and T.O. were responsible for project development, data collection, and data analysis; K.K. was responsible for data collection; N.M. were responsible for project development and supervision. All authors contributed to reviewing and editing the manuscript. All authors have read and agreed to the published version of the manuscript.

Funding: This work was supported by an unrestricted research grant from the Ibaraki Prefecture (91AA193108).

Institutional Review Board Statement: The research protocol was reviewed and approved by the Tokyo Medical and Dental University Review Board (number: 774, approval date: 23 March 2010), and written informed consent was obtained from all participants. This study was conducted following the Declaration of Helsinki.

Informed Consent Statement: Informed consent was obtained from all subjects involved in the study.

Acknowledgments: M.T. reports research grants from Ibaraki Prefecture and personal fees for the speaker's bureau from Fuji Pharma Co Ltd.

Conflicts of Interest: The authors declare no conflict of interest. The funders had no role in the design of the study; in the collection, analyses, or interpretation of data; in the writing of the manuscript; or in the decision to publish the results.

References

1. Lauritsen, C.G.; Chua, A.L.; Nahas, S.J. Current Treatment Options: Headache Related to Menopause-Diagnosis and Management. *Curr. Treat. Options Neurol.* **2018**, *20*, 7. [CrossRef] [PubMed]
2. Terauchi, M.; Hiramitsu, S.; Akiyoshi, M.; Owa, Y.; Kato, K.; Obayashi, S.; Matsushima, E.; Kubota, T. Effects of the kampo formula tokishakuyakusan on headaches and concomitant depression in middle-aged women. *Evid.-Based Complementary Altern. Med. Ecam* **2014**, *2014*, 593560. [CrossRef] [PubMed]
3. Lieba-Samal, D.; Wober, C. Sex hormones and primary headaches other than migraine. *Curr. Pain Headache Rep.* **2011**, *15*, 407–414. [CrossRef] [PubMed]
4. Ripa, P.; Ornello, R.; Degan, D.; Tiseo, C.; Stewart, J.; Pistoia, F.; Carolei, A.; Sacco, S. Migraine in menopausal women: A systematic review. *Int. J. Women's Health* **2015**, *7*, 773–782. [CrossRef]
5. Razeghi Jahromi, S.; Ghorbani, Z.; Martelletti, P.; Lampl, C.; Togha, M.; EHF-SAS. Association of diet and headache. *J. Headache Pain* **2019**, *20*, 106. [CrossRef]
6. Ramsden, C.E.; Zamora, D.; Faurot, K.R.; Macintosh, B.; Horowitz, M.; Keyes, G.S.; Yuan, Z.-X.; Miller, V.; Lynch, C.; Honvoh, G.; et al. Dietary alteration of n-3 and n-6 fatty acids for headache reduction in adults with migraine: Randomized controlled trial. *BMJ* **2021**, *374*, n1448. [CrossRef]
7. Odai, T.; Terauchi, M.; Suzuki, R.; Kato, K.; Hirose, A.; Miyasaka, N. Severity of subjective forgetfulness is associated with high dietary intake of copper in Japanese senior women: A cross-sectional study. *Food Sci. Nutr.* **2020**, *8*, 4422–4431. [CrossRef]
8. Kobayashi, S.; Honda, S.; Murakami, K.; Sasaki, S.; Okubo, H.; Hirota, N.; Notsu, A.; Fukui, M.; Date, C. Both comprehensive and brief self-administered diet history questionnaires satisfactorily rank nutrient intakes in Japanese adults. *J. Epidemiol.* **2012**, *22*, 151–159. [CrossRef]
9. Kobayashi, S.; Murakami, K.; Sasaki, S.; Okubo, H.; Hirota, N.; Notsu, A.; Fukui, M.; Date, C. Comparison of relative validity of food group intakes estimated by comprehensive and brief-type self-administered diet history questionnaires against 16 d dietary records in Japanese adults. *Public Health Nutr.* **2011**, *14*, 1200–1211. [CrossRef]
10. Hanson, B.S.; Isacsson, S.O.; Janzon, L.; Lindell, S.E. Social network and social support influence mortality in elderly men. The prospective population study of "Men born in 1914," Malmö, Sweden. *Am. J. Epidemiol.* **1989**, *130*, 100–111. [CrossRef]
11. Hunter, M.; Battersby, R.; Whitehead, M. Relationships between psychological symptoms, somatic complaints and menopausal status. *Maturitas* **1986**, *8*, 217–228. [CrossRef]
12. van Keep, P.A.; Kellerhals, J.M. The impact of socio-cultural factors on symptom formation. Some results of a study on ageing women in Switzerland. *Psychother. Psychosom.* **1974**, *23*, 251–263. [CrossRef] [PubMed]
13. Zigmond, A.S.; Snaith, R.P. The hospital anxiety and depression scale. *Acta Psychiatr. Scand.* **1983**, *67*, 361–370. [CrossRef] [PubMed]
14. Gomez-Zorita, S.; Gonzalez-Arceo, M.; Fernandez-Quintela, A.; Eseberri, I.; Trepiana, J.; Portillo, M.P. Scientific Evidence Supporting the Beneficial Effects of Isoflavones on Human Health. *Nutrients* **2020**, *12*, 3853. [CrossRef]
15. Wei, H.; Bowen, R.; Cai, Q.; Barnes, S.; Wang, Y. Antioxidant and antipromotional effects of the soybean isoflavone genistein. *Proc. Soc. Exp. Biol. Med. Soc. Exp. Biol. Med.* **1995**, *208*, 124–130. [CrossRef]
16. Chen, L.R.; Chen, K.H. Utilization of Isoflavones in Soybeans for Women with Menopausal Syndrome: An Overview. *Int. J. Mol. Sci.* **2021**, *22*, 3212. [CrossRef]
17. Perna, S.; Peroni, G.; Miccono, A.; Riva, A.; Morazzoni, P.; Allegrini, P.; Preda, S.; Baldiraghi, V.; Guido, D.; Rondanelli, M. Multidimensional Effects of Soy Isoflavone by Food or Supplements in Menopause Women: A Systematic Review and Bibliometric Analysis. *Nat. Prod. Commun.* **2016**, *11*, 1733–1740. [CrossRef]
18. Glazier, M.G.; Bowman, M.A. A review of the evidence for the use of phytoestrogens as a replacement for traditional estrogen replacement therapy. *Arch. Intern. Med.* **2001**, *161*, 1161–1172. [CrossRef]
19. Messina, M.J. Legumes and soybeans: Overview of their nutritional profiles and health effects. *Am. J. Clin. Nutr.* **1999**, *70*, 439S–450S. [CrossRef]
20. Allais, G.; Castagnoli Gabellari, I.; Burzio, C.; Rolando, S.; De Lorenzo, C.; Mana, O.; Benedetto, C. Premenstrual syndrome and migraine. *Neurol. Sci. Off. J. Ital. Neurol. Soc. Ital. Soc. Clin. Neurophysiol.* **2012**, *33* (Suppl. S1), 111. [CrossRef]
21. Bryant, M.; Cassidy, A.; Hill, C.; Powell, J.; Talbot, D.; Dye, L. Effect of consumption of soy isoflavones on behavioural, somatic and affective symptoms in women with premenstrual syndrome. *Br. J. Nutr.* **2005**, *93*, 731–739. [CrossRef] [PubMed]

22. Guo, P.P.; Li, P.; Zhang, X.H.; Liu, N.; Wang, J.; Chen, D.D.; Sun, W.J.; Zhang, W. Complementary and alternative medicine for natural and treatment-induced vasomotor symptoms: An overview of systematic reviews and meta-analyses. *Complementary Ther. Clin. Pract.* **2019**, *36*, 181–194. [CrossRef] [PubMed]
23. Jacobs, A.; Wegewitz, U.; Sommerfeld, C.; Grossklaus, R.; Lampen, A. Efficacy of isoflavones in relieving vasomotor menopausal symptoms-A systematic review. *Mol. Nutr. Food Res.* **2009**, *53*, 1084–1097. [CrossRef] [PubMed]
24. Rapkin, A.J. Vasomotor symptoms in menopause: Physiologic condition and central nervous system approaches to treatment. *Am. J. Obstet. Gynecol.* **2007**, *196*, 97–106. [CrossRef] [PubMed]
25. Hirose, A.; Terauchi, M.; Akiyoshi, M.; Owa, Y.; Kato, K.; Kubota, T. Low-dose isoflavone aglycone alleviates psychological symptoms of menopause in Japanese women: A randomized, double-blind, placebo-controlled study. *Arch. Gynecol. Obstet.* **2016**, *293*, 609–615. [CrossRef]
26. Fernandez-de-Las-Penas, C.; Fernandez-Munoz, J.J.; Palacios-Cena, M.; Paras-Bravo, P.; Cigaran-Mendez, M.; Navarro-Pardo, E. Sleep disturbances in tension-type headache and migraine. *Ther. Adv. Neurol. Disord.* **2017**, *11*. [CrossRef]
27. Minen, M.T.; Begasse De Dhaem, O.; Kroon Van Diest, A.; Powers, S.; Schwedt, T.J.; Lipton, R.; Silbersweig, D. Migraine and its psychiatric comorbidities. *J. Neurol. Neurosurg. Psychiatry* **2016**, *87*, 741–749. [CrossRef]
28. Messina, M. Soy and Health Update: Evaluation of the Clinical and Epidemiologic Literature. *Nutrients* **2016**, *8*, 754. [CrossRef]
29. Burger, H.G.; Dudley, E.C.; Hopper, J.L.; Groome, N.; Guthrie, J.R.; Green, A.; Dennerstein, L. Prospectively measured levels of serum follicle-stimulating hormone, estradiol, and the dimeric inhibins during the menopausal transition in a population-based cohort of women. *J. Clin. Endocrinol. Metab.* **1999**, *84*, 4025–4030. [CrossRef]
30. MacGregor, E.A. Migraine headache in perimenopausal and menopausal women. *Curr. Pain Headache Rep.* **2009**, *13*, 399–403. [CrossRef]
31. Chai, N.C.; Peterlin, B.L.; Calhoun, A.H. Migraine and estrogen. *Curr. Opin. Neurol.* **2014**, *27*, 315–324. [CrossRef] [PubMed]
32. Allais, G.; Chiarle, G.; Sinigaglia, S.; Airola, G.; Schiapparelli, P.; Benedetto, C. Estrogen, migraine, and vascular risk. *Neurol. Sci. Off. J. Ital. Neurol. Soc. Ital. Soc. Clin. Neurophysiol.* **2018**, *39*, 11–20. [CrossRef] [PubMed]
33. Jia, M.; Dahlman-Wright, K.; Gustafsson, J.A. Estrogen receptor alpha and beta in health and disease. *Best Pract. Res. Clin. Endocrinol. Metab.* **2015**, *29*, 557–568. [CrossRef] [PubMed]
34. Reed, S.D.; Lampe, J.W.; Qu, C.; Gundersen, G.; Fuller, S.; Copeland, W.K.; Newton, K.M. Self-reported menopausal symptoms in a racially diverse population and soy food consumption. *Maturitas* **2013**, *75*, 152–158. [CrossRef] [PubMed]

MDPI
St. Alban-Anlage 66
4052 Basel
Switzerland
Tel. +41 61 683 77 34
Fax +41 61 302 89 18
www.mdpi.com

Nutrients Editorial Office
E-mail: nutrients@mdpi.com
www.mdpi.com/journal/nutrients

www.ingramcontent.com/pod-product-compliance
Lightning Source LLC
LaVergne TN
LVHW070840170726
843515LV00005B/536